The Logic of
Biochemical Sequencing

The Logic of Biochemical Sequencing

David S. Blackman, Ph.D.
Professor of Chemistry
Department of Chemistry
University of the District of Columbia
Washington, DC

CRC Press
Boca Raton Ann Arbor London Tokyo

Library of Congress Cataloging-in-Publication Data

Blackman, David.
The logic of biochemical sequencing / David Blackman.
p. cm.
Includes bibliographical references (p.) and index.
ISBN 0-8493-4497-2
1. Amino acid sequence—Philosophy. 2. Logic. 3. Analysis (Philosophy) I. Title.
QP551.B53 1993
574.87'328—dc 20 93-12778
CIP

International Standard Book Number 0-8493-4497-2
Library of Congress Card Number 93-12778
Printed in the United States of America 1 2 3 4 5 6 7 8 9 0
Printed on acid-free paper

Dedication

For Susan and Elaine, and the real Sam, Michael, and Rayna, each of whom has, at one time or another, put up with more from me than I had any right to expect.

Acknowledgments

I have often been bemused by the lists of names presented at the beginnings of books. It seemed, in my naïveté, that a book is the sole product of the author. That such is not true became abundantly evident as I labored on this work. My colleagues at the University of the District of Columbia have provided assistance and emotional support for which I can only begin to express my thanks. I mention them in no particular order, in the hope that they will appreciate how much I am indebted to Debbie Whalen, Joyce Butler, Van Van Nguyen, Charles Ester, George Eng, Muriel Prouty, Hershel McDowell, and Pat Thorstenson. Jeff Holtmeier and Monique Power at CRC Press have provided invaluable advice on bringing this work to fruition. My editor at CRC, John Herring, brought excitement and a healthy dose of good-natured cynicism to this project. I am deeply grateful for his guidance, and for his having given me the opportunity to share a lovely literary brunch while escaping the hot Florida sun.

Much of this book was written during the summers of 1990 and 1991, under the terms of Summer Research Grants awarded by the Faculty Senate of the University of the District of Columbia, whose support is gratefully acknowledged.

Sources

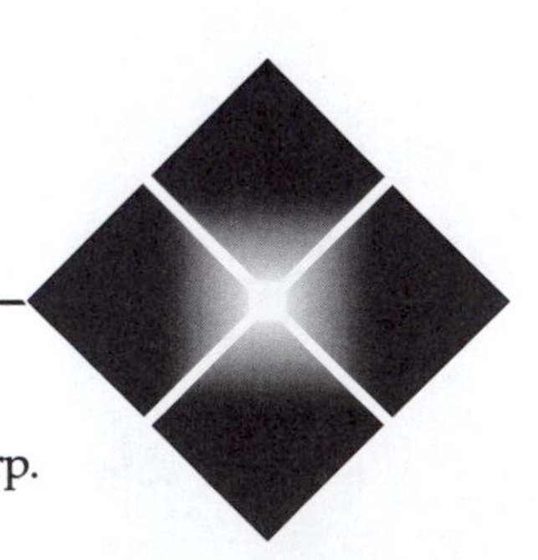

pGEM® is a registered trademark of Promega Corp.
The 1 Kb Ladder is a product of GIBCO-BRL.

Preface

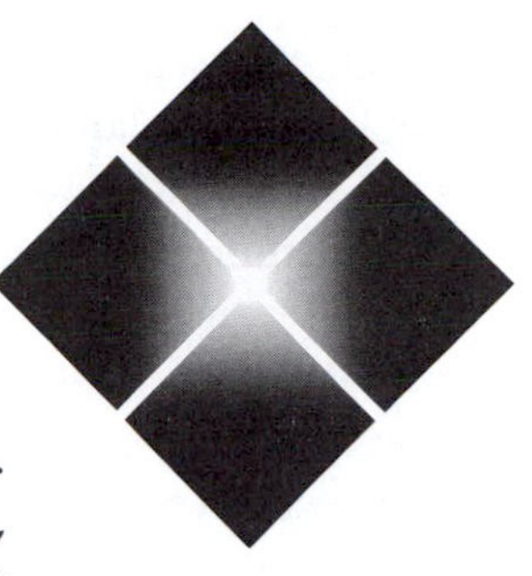

A story is told of two men examining a portrait. When one man asks who the subject of the painting is, the other replies, "Brothers and sisters have I none, but that man's father is my father's son."

One of my former professors once noted that the people of the world are divided into two groups: those who love the work of the Dutch artist M. C. Escher and those who hate it. My contention is that the dichotomy of the world is between those who love logical puzzles and those who hate them. (It is, of course, possible that we are both right, and furthermore, that the Escher lovers and logic lovers are the same people.) Both groups can benefit from the study of biochemical sequencing problems.

In the strictest sense, this book is not about sequencing in biochemistry. That is to say, it is neither a laboratory manual nor a monograph describing sequencing techniques and methods. What it is about is the thought processes needed to understand and interpret the results of sequencing experiments. Very few of the users of this book will probably ever carry out any sequencing experiments. All of them, however, will ultimately be called upon to exercise judgments based on logical arguments, and logic is the keystone around which this book is built.

It is not necessary to understand a great deal of biochemistry or molecular biology to make use of the book. After all, in my own course, we do protein sequencing problems early on in the first semester. The "methods" described here provide the logical groundwork for solving sequencing problems. I have made no attempt to provide any information on the actual experimental procedures. These are available elsewhere. My hope is that, after studying the problems presented here, it will be possible for the student to understand and, if necessary, carry out sequencing experiments. More immediately, I hope that the reader will come to appreciate the amount of time and thought needed to devise and understand such work. But most importantly, I hope that study of this book will guide my readers to a more extensive use of logic in everyday life. If use of the examples presented here helps the reader to improve on those thought processes, this book will have achieved its goal.

This book is meant to be read sequentially (if you will pardon the pun). Each of the first four chapters begins with a brief discussion of the theory of one aspect of sequencing, and presents several illustrative examples of increasing complexity. End of chapter problems test your mastery of the material. Chapter 5 provides several more problems. Completely worked out solutions to the end-of-chapter problems are found in Appendix A. The answers to the problems in Chapter 5 are presented without detail in Appendix B. Any time you feel the urge for a little escapism you are invited to read the short stories in Chapter 6. These are based on the sequencing principles presented in the text, and each story contains sufficient clues to solve the problem it contains. Complete solutions are provided, but you are urged first to try the problems on your own.

Welcome to the wonderful world of sequencing and to the joy of logical thinking. Whether you recognized immediately that the portrait was of the second man's son, or you required some time to reason it out, you have already begun to use the kind of thought processes needed to solve the logical puzzles presented here.

David Blackman, Ph.D.

The Author

David Blackman, Ph.D., is Professor of Chemistry at the University of the District of Columbia, Washington, D.C.

Dr. Blackman earned his B.S. and M.A. in chemistry from Brooklyn Collge, City University of New York, in 1962 and 1965, respectively, and a Ph.D. in biochemistry from Columbia University, New York, in 1971. He has participated in NSF-Chautauqua short courses at the University of Maryland, the University of Puerto Rico, and Temple University.

Dr. Blackman has also taught at Brooklyn College Brooklyn, New York; New York Institute of Technology, New York, New York; Fairleigh Dickinson University, Teaneck, New Jersey; and Antioch College, Columbia, Maryland. He has held his present position since 1971.

Dr. Blackman is a member of the American Chemical Society, the Chemical Society of Washington, the honorary society Sigma Xi, and the Beta Kappa Chi national scientific honor society. He was the recipient of a MARC Faculty Fellowship from the National Institutes of Health.

Dr. Blackman has been a guest worker in the Laboratory of Chemical Biology of the National Institute of Arthritis, Metabolism, and Digestive Diseases, as well as in the Laboratory of Microbiology and Immunology of the National Institue of Dental Research, both at the National Institutes of Health in Bethesda, Maryland. His research has been supported by the National Institutes of Health, the United States Department of Agriculture, and the United States Department of the Interior.

Dr. Blackman has published several articles in *Biochemical Education* and in the *Journal of Chemical Education*, and has presented papers at national and regional meetings. Among his current research interests is the biochemical characterization of *Hydrilla verticillata*.

Contents

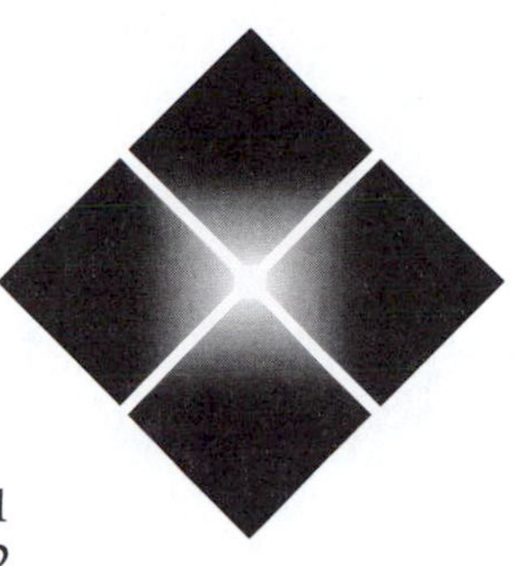

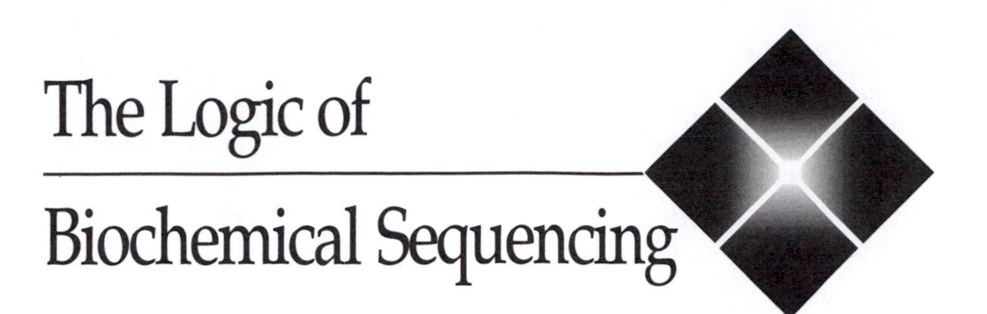
The Logic of
Biochemical Sequencing

End Group Analysis of Proteins

Introduction

For purposes of sequencing experiments, a protein can be considered as nothing more than a string of amino acids joined by peptide bonds, that is, a polypeptide. Higher orders of structure are irrelevant; in fact, they are often destroyed experimentally prior to carrying out any cleavage reactions. The only exception is the disulfide bond which, as we will see later, contributes to our understanding of primary structure.

The general outline of protein sequencing problems involves the following steps:

1. Destroy disulfide bonds by reduction.
2. Determine the amino acid composition of the protein or peptide. This requires complete hydrolysis, which is accomplished by heating for 22 h at 120°C in 6N HCl followed by amino acid analysis. The amino acid analyses presented in this book will contain molar ratios of products and will be in whole numbers. In a real experiment, of course, the values that are obtained will usually be fractional and will depend on the amount of material submitted for analysis. It is important to keep in mind the fact that hydrolysis can change the character of an amino acid. While we will not be concerned with degradation of amino acids such as tryptophan and methionine (both of which will always be reported as the number of residues in the original peptide), asparagine and glutamine will, of necessity, be reported as aspartate and glutamate, reflecting the fact that peptide hydrolysis also hydrolyzes amides. Consequently, unless you are told otherwise, assume that any Asp or Glu residues reported in the amino acid analysis represent either the amino acid or the corresponding amide. You should also be aware that *N*-terminal blocking groups, e.g., acetyl, formyl, etc., while relatively rare, will also be hydrolyzed.

3. Determine the amino and carboxyl termini.
4. Starting with a sample of intact, reduced peptide, use one of several cleavage reactions. Generally this will yield a mixture of smaller peptide fragments. In a laboratory situation these would be resolved by chromatographic or electrophoretic methods, and each product would then be subjected to complete hydrolysis and amino acid analysis. For the simulations in this book, the purification steps will be assumed and amino acid analyses will be reported in order of increasing fragment size.
5. Extract as much information as possible from each of the fragments. Based on the known specificity of the cleavage method it is possible to determine at least one of the termini of a given fragment; you may also be able to locate one or more of these fragments in the original protein. Sometimes additional data will be needed, such as an experiment to determine the carboxyl or amino terminus of a fragment, its behavior in an electrophoresis experiment, or the results of a second cleavage reaction. In any case, it is crucial that you deduce as much sequence information as possible from each experiment, since there are usually *no* redundant data. That is to say, each problem will contain just enough information to solve the sequence, assuming you can extract all the information available.
6. In virtually every case it will be necessary to utilize at least two different cleavage reactions to determine the entire sequence. You will have to exercise a certain amount of ingenuity as you go back and forth among experiments, transferring conclusions from one to another. Often this will feel like watching a tennis match, following the ball from one side of the court to the other. The best advice is to maintain good books, keeping track of peptide fragments and experimental conclusions until, finally, the entire sequence falls out.
7. If cysteine residues are found, the possibility of disulfide bonds should be considered. In that case the original, unreduced peptide must be subjected to at least one cleavage so as to confirm the presence, and determine the location, of the disulfide linkage.

Reactions of the Amino Group

The amino terminus of a peptide may be found by any number of methods. Historically, Sanger's reagent, 1-fluoro-2,4-dinitrobenzene (FDNB), was the earliest reagent used. Recently, however, this has been replaced by "dansyl" chloride (1-dimethylaminonaphthalene-5-sulfonyl chloride), which yields highly fluorescent *N*-dansylamino acids, and is therefore more sensitive by a factor of about 100. Conceptually, however, what is said about the use of FDNB applies equally to dansyl chloride.

A third *N*-terminal reagent, phenylisothiocyanate (PITC or Edman's reagent) has the advantage of derivatizing the *N*-terminal amino group, cleaving it as the phenylthiohydantoin (PTH), and leaving the remainder of the peptide intact. In principle, the Edman method could be applied sequentially to determine the entire sequence of a peptide, working from the amino toward the carboxyl end. This has been done, although after about 50 or so residues the results begin to get muddy. Nonetheless, a great deal of time can be saved by using the sequential Edman procedure in an automatic instrument called a sequenator. While this instrument has made it possible to deduce the sequence of a fairly hefty peptide in a very short time, its use detracts from the goal of teaching logical thought processes. As was noted in the preface, it is not the aim of this book to teach the actual techniques of sequencing — it is to help you learn to solve problems by thinking logically. In fact, as we go through the problems in this book, the Edman reaction will rarely be used.

A serious potential problem with the chemical methods is that the side chain amino group of lysine might react in the same way as the *N*-terminal amino group. This can generally be avoided by maintaining the pH of the reaction medium so that the α-amino group is unprotonated and reactive, while the ε-amino group is protonated and unreactive. Fortunately this is quite easy to do, since the pK values of the two groups are sufficiently different. In the case of the Sanger (FDNB) procedure, for example, adding bicarbonate to the reaction mixture maintains the pH at the appropriate value.

After derivatization of the α-amino group the peptide is hydrolyzed, and the hydrolyzate is resolved by chromatography on paper for 2,4-dinitrophenyl (DNP) or dansyl derivatives, or by gas/liquid chromatography for PTH derivatives. Paper chromatography in the presence of known standards will, eventually, yield the identity of the yellow (DNP) or fluorescent (dansyl) derivative. A GLC column must first be standardized by chromatography of known PTH amino acids.

Example 1.1

Hydrolysis and amino acid analysis of a dipeptide yields only alanine. What is its amino acid sequence?

In general, the amino acid sequence of any homopeptide (i.e., a peptide containing only one type of amino acid) is immediately defined (but see Example 1.2). Since both positions in the dipeptide contain alanine, the sequence must be Ala–Ala.

Example 1.2

A dipeptide, after hydrolysis and amino acid analysis, yields only glutamic acid. What is its amino acid sequence?

Finding either glutamic or aspartic acid in an amino acid analysis should always alert you to the possibility of glutamine or asparagine in the original peptide. Subsequent data often clarify the situation. It is useful to indicate Glu/Gln or Asp/Asn uncertainty with the abbreviation "Glx" or "Asx".

In this case there are four possible dipeptides, namely Glu–Glu, Glu–Gln, Gln–Glu, or Gln–Gln. More succinctly we can say that the sequence is Glx–Glx.

Example 1.3

A dipeptide contains 1 residue each of tyrosine and valine. Treatment of the dipeptide with FDNB followed by hydrolysis gives DNP–valine. What is the amino acid sequence of the peptide?

There are only two possible sequences for a dipeptide containing two different amino acids. In this case the sequence must be either Tyr–Val or Val–Tyr. If either the amino or carboxyl terminal amino acid is known, the sequence is defined, since the other amino acid must be at the opposite end. Here the amino terminus is valine; thus the carboxyl terminus is tyrosine, and the sequence is Val–Tyr.

Example 1.4

The amino acid composition of an unknown tripeptide is (Ala_2,Arg). Treatment of the peptide with FDNB followed by hydrolysis gives DNP–arginine. What is the amino acid sequence of the tripeptide?

The amino terminal amino acid is arginine. Both remaining positions are occupied by alanine. Thus the sequence must be Arg–Ala–Ala. Note that it is not proper to record this sequence as "Arg–Ala_2". When the sequence is known, each amino acid should be noted in its proper position.

Example 1.5

The amino acid composition of an unknown tripeptide is (Ala_2,Arg). Treatment of the peptide with FDNB followed by hydrolysis gives DNP–alanine. What is the amino acid sequence of the tripeptide?

The amino terminus is alanine, but the remaining two positions are as yet undefined. The sequence could be either Ala–Ala–Arg or Ala–Arg–Ala. A convenient way to represent an unknown sequence is to list the amino acids in parentheses separated by commas, as in Ala–(Ala,Arg). Additional data are needed to solve this problem.

Example 1.6

The amino acid composition of an unknown tripeptide is (Ala_2,Arg). Treatment of the peptide with FDNB followed by hydrolysis gives no DNP amino acids. What can be said about the amino acid sequence of this peptide?

Reaction with FDNB requires a free, unprotonated amino group. Any amino group that is protonated (hence positively charged), e.g., the ε-amino group of lysine, or is derivatized with, e.g., a formyl or an acetyl group, will not react. Since it is assumed that the dinitrophenylation reaction was carried out, as usual, in the presence of a bicarbonate buffer, dinitrophenylation will be confined to the free α-amino group. Hence the data suggest blockage of the *N*-terminal amino group. It is also possible that the peptide is actually cyclic, with the *C*-terminal carboxyl group bound to the *N*-terminal amino group. At this point, nothing can be said about the sequence of this peptide.

Example 1.7

The amino acid composition of an unknown tripeptide is (Glu,Ser,Tyr). Treatment of the peptide with FDNB followed by hydrolysis gives DNP–serine. What can be said about the amino acid sequence of this peptide?

The *N*-terminal amino acid is clearly serine; the two remaining positions cannot yet be established. Note, however, that there are four (rather than two) possible sequences. As we saw in Example 1.2, whenever glutamic or aspartic acid appears in the amino acid analysis, the possibility should be considered that the amino acid might have been glutamine or asparagine. Side chain amide groups, of course, will be lost during hydrolysis. The possible sequences are Ser–Glu–Tyr, Ser–Gln–Tyr, Ser–Tyr–Glu, and Ser–Tyr–Gln, or more succinctly, Ser–(Glx,Tyr).

The problem of distinguishing between the acidic amino acids and their amides will be discussed in the next section and in Chapter 2.

Reactions of the Carboxyl Group

At first glance it would seem that determination of the carboxyl terminus of a peptide would be straightforward: simply treat the peptide with carboxypeptidase, isolate the single amino acid product, and identify it. There are, in fact, two analytically important carboxypeptidases with differing specificities. Carboxypeptidase A catalyzes the hydrolysis of all *C*-terminal amino acids with the exception of proline; carboxypeptidase B is specific for arginine and lysine. Neither enzyme will remove the *C*-terminal amino acid following a prolyl

Table 1
Lithium Borohydride Reduction Products of the *C*-Terminal Amino Acids

Amino acid	Reduction product
Alanine	2-amino-1-propanol
Arginine	2-amino-5-guanido-1-pentanol
Asparagine	3-amino-4-hydroxybutanamide[a]
Aspartic acid	2-amino-1,4-butanediol
Cysteine	2-amino-3-mercapto-1-propanol
Glutamic acid	2-amino-1,5-pentanediol
Glutamine	4-amino-5-hydroxypentanamide[b]
Glycine	2-aminoethanol
Histidine	2-amino-3-imidazolyl-1-propanol
Isoleucine	2-amino-3-methyl-1-pentanol
Leucine	2-amino-4-methyl-1-pentanol
Lysine	2,6-diamino-1-hexanol
Methionine	2-amino-4-(methylthio)-1-butanol
Phenylalanine	2-amino-3-phenyl-1-propanol
Proline	2-hydroxymethylpyrrolidine
Serine	2-amino-1,3-propanediol
Threonine	2-amino-1,3-butanediol
Tryptophan	2-amino-3-indolyl-1-propanol
Tyrosine	2-amino-3-(*p*-hydroxyphenyl)-1-propanol
Valine	2-amino-3-methyl-1-butanol

[a] Subsequent hydrolysis yields 3-amino-4-hydroxybutanoic acid.
[b] Subsequent hydrolysis yields 4-amino-5-hydroxypentanoic acid.

residue. In practice, one would use a mixture of carboxypeptidases A and B to cleave the *C*-terminus. As long as proline is not encountered, hydrolysis is expected to proceed smoothly. The difficulty, of course, is that as soon as the *C*-terminal amino acid is removed, the enzyme "sees" a new *C*-terminus, and begins to act on it. As a result, several different amino acids may appear in the product mixture, and the identity of the true *C*-terminal amino acid may not be immediately apparent. Note, of course, that the same problem exists in using leucine aminopeptidase to identify the *N*-terminal amino acid.

A much less ambiguous (and, for our purposes, much more interesting) reagent is lithium borohydride. Treatment of a peptide with this reducing agent converts all carboxyl groups to hydroxymethyl ($-CH_2OH$) groups. The products of hydrolysis of the reduced peptide will have altered ionic properties, since carboxyl groups are negatively charged at neutral or alkaline pH while hydroxymethyl groups are uncharged. The amino alcohols can be separated and identified by ion exchange chromatography in the amino acid analyzer, and much information can be obtained with regard to peptide structure.

As an exercise, show the structures and derive the names of the borohydride reduction products of all the amino acids. These are listed in Table 1, but it is strongly suggested that you solve this problem by yourself.

Example 1.8

A tripeptide has the composition (Gly_2,Lys). Treatment with lithium borohydride followed by hydrolysis gives 2,6-diamino-1-hexanol as the only product lacking a carboxyl group. What is the sequence of this tripeptide?

The reduction product arose from lysine, which must have been at the *C*-terminus. The two glycine residues come from positions 1 and 2, and the sequence is Gly–Gly–Lys.

Example 1.9

A tripeptide has the composition (Gly_2,Lys). Treatment with lithium borohydride followed by hydrolysis gives 2-aminoethanol as the only product lacking a carboxyl group. What is the sequence of this tripeptide?

The *C*-terminal amino acid is glycine. Positions 1 and 2 cannot be determined without additional information. The sequence can be shown as (Gly,Lys)–Gly.

Example 1.10

A dipeptide has the amino acid composition (Asp,Ser). Treatment with lithium borohydride followed by hydrolysis gives 2-amino-1,4-butanediol and serine. What is the amino acid sequence of the dipeptide?

Reduction of *C*-terminal aspartic acid yields 2-amino-1,4-butanediol; thus, the amino acid sequence is Ser–Asp.

Example 1.11

A dipeptide has the amino acid composition (Asp,Ser). Treatment with lithium borohydride followed by hydrolysis gives 3-amino-4-hydroxybutanoic acid and serine. What is the amino acid sequence of the dipeptide?

Asparagine at the *C*-terminus will be reduced to 3-amino-4-hydroxybutanamide, but subsequent hydrolysis will convert this to the corresponding acid, 3-amino-4-hydroxybutanoic acid. Thus the sequence is Ser–Asn. Note how the reduction/hydrolysis product of aspartic acid differs from that of asparagine.

Example 1.12

A dipeptide has the amino acid composition (Asp,Ser). Treatment with lithium borohydride followed by hydrolysis gives 2-amino-1,3-propanediol and 2-amino-4-hydroxybutanoic acid. What is the amino acid sequence of the dipeptide?

The first product, 2-amino-1,3-propanediol, clearly arises from *C*-terminal serine. But treatment of a peptide with lithium borohydride reduces *all* carboxyl groups to hydroxymethyl groups. Aspartic acid at other than the *C*-terminus yields 2-amino-4-hydroxybutanoic acid. (Compare this to the product obtained from *C*-terminal asparagine in the preceding example.) The sequence is Asp–Ser.

Example 1.13

When a dipeptide with the amino acid composition (Asp,Ser) is reduced with lithium borohydride and hydrolyzed, 2-amino-1,3-propanediol and aspartic acid are obtained. What is the amino acid sequence of the dipeptide?

C-terminal serine yields 2-amino-1,3-propanediol. Aspartic acid, then, must come from the *N*-terminus. The only way this could happen is to have *N*-terminal asparagine, since *N*-terminal aspartic acid would yield 2-amino-4-hydroxybutanoic acid. The sequence is Asn–Ser.

Example 1.14

A tripeptide has the amino acid composition (Lys,Tyr,Val). Treatment with FDNB followed by hydrolysis gives DNP–tyrosine. Lithium borohydride reduction followed by hydrolysis gives 2,6-diamino-1-hexanol. What is the amino acid sequence?

In general, when two procedures are carried out on the same material, they are done independently. That is, two separate samples are used; the material treated with FDNB is discarded after it is analyzed, and another sample of the original tripeptide is subsequently treated with lithium borohydride.

The FDNB data indicate *N*-terminal tyrosine, while the lithium borohydride data indicate *C*-terminal lysine. The sequence is Tyr–Val–Lys. Note that for a tripeptide the entire sequence is established when the *N*- and *C*-termini are known.

Example 1.15

A tetrapeptide has the amino acid composition (His,Ile,Leu,Phe). Lithium borohydride reduction followed by hydrolysis gives 2-amino-3-methyl-1-pentanol. Dinitrophenylation followed by hydrolysis gives DNP–histidine. What is the amino acid sequence?

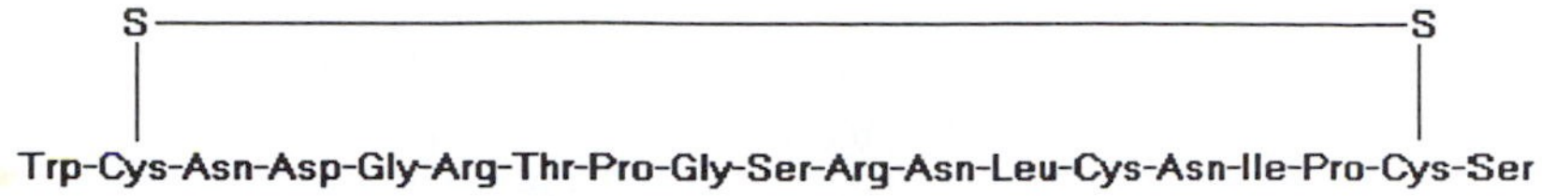

S———————————————————————S

Trp-Cys-Asn-Asp-Gly-Arg-Thr-Pro-Gly-Ser-Arg-Asn-Leu-Cys-Asn-Ile-Pro-Cys-Ser

FIGURE 1 Peptide sequence for Problem 1.1E.

The terminal amino acids can be identified, but those in positions 2 and 3 cannot. A partial sequence can be written as His–(Leu,Phe)–Ile. Note again the use of parentheses and commas to indicate unknown sequences.

Example 1.16

A tetrapeptide has the amino acid composition (His,Ile,Leu,Phe). Treatment with Edman's reagent gives PTH–histidine and a tripeptide. Dinitrophenylation of this tripeptide followed by hydrolysis gives DNP-leucine. Treatment of either the tripeptide or the original tetrapeptide with lithium borohydride followed by hydrolysis gives 2-amino-3-methyl-1-pentanol. What is the amino acid sequence?

Note that this problem is similar to the previous one, except that the information now presented is sufficient to determine the entire sequence. The Edman reaction converts the *N*-terminal amino acid to its PTH (phenylthiohydantoin) derivative, leaving the remainder of the original peptide intact. The problem now reduces to a simple tripeptide determination. The entire sequence is His–Leu–Phe–Ile.

End of Chapter Problems

Problem 1.1

Predict the results of amino acid analyses carried out on hydrolyzates of the following peptides:

A. Leu–Val–Thr–Leu–Ala–Cys–His–His (residues 106 to 113 of the α chain of dog hemoglobin)

B. Leu–Arg–Val–Asp–Pro–Val–Asn–Phe–Lys (residues 91 to 99 of the hemoglobin α chain of the gray kangaroo)

C. Lys–Pro–Leu–Ala–Gln–Ser–His–Ala–Thr–Lys–His (residues 87 to 97 of dolphin myoglobin)

D. Lys–Gln–Thr–Ile–Ala–Ser–Asn (the *C*-terminal heptapeptide of bovine trypsinogen)

E. The peptide in Figure 1 (residues 63 to 81 of chicken lysozyme; residues 64 and 80 are joined by a disulfide bond)

F. His–Ser–Asp–Gly–Thr–Phe–Thr–Ser–Glu–Leu–Ser–Arg–Leu–Arg–Asp–Ser–Ala–Arg–Leu–Gln–Arg–Leu–Leu–Gln–Gly–Leu–Val–NH_2 (pig secretin, the *C*-terminal valine is amidated)

Problem 1.2

There are four possible dipeptides that can be made from aspartic acid and asparagine. For each of them, predict the products of lithium borohydride reduction followed by hydrolysis.

Problem 1.3

In each of the following cases assume that a single sample of the peptide is treated first with FDNB, then with lithium borohydride, and is finally hydrolyzed and subjected to amino acid analysis. List the products that would be obtained.

A. Ala–Gln–Ser–Val–Pro–Tyr–Gly (the *N*-terminal heptapeptide of *Bacillus amyloliquefaciens* subtilisin)

B. Ser–Asn–Ala–Cys–Lys–Asn–His–Gln–Arg–Phe (residues 118 to 127 of bovine rennin)

C. Asn–Ala–Asn–Val–Leu–Ala–Arg–Tyr–Ala–Ser–Ile–Cys (residues 24 to 35 of rabbit muscle aldolase A)

D. Lys–Glu–Leu–Asn–Asp–Leu–Glu–Lys–Lys–Tyr–Asn (residues 1 to 11 of *Streptococcus aureus* PC-1 penicillinase)

E. Asn–Ile–Leu–Gly–Arg–Glu–Ala–Lys–Cys–Thr–Asn–Glu (residues 1 to 12 of bovine pancreatic secretory trypsin inhibitor)

F. Gly–Tyr–Ile–Val–Asp–Asp–Lys–Asn–Cys–Thr (residues 4 to 13 of scorpion neurotoxin V)

G. Ser–Ala–Phe–Arg–Asn–Glu–Glu–Tyr–Asn–Lys–Ser (residues 37 to 47 of human orosomucoid)

H. Lys–Glu–Arg–Gly–Ala–Leu–Pro–Met–Ile–Asp–Arg–Gly–Asp–Ile–Arg–Gln (residues 99 to 114 of bacteriophage λ endolysin)

I. Ala–His–Gly–Lys–Lys–Val–Ile–Thr–Ala–Phe–Ser–Asp–Gly–Leu (residues 62 to 75 of mouse hemoglobin β chain)

J. Leu–Ala–Lys–Val–Ile–His–Asp–His–Phe–Gly–Ile (residues 157 to 167 of pig glyceraldehyde-3-phosphate dehydrogenase)

Limited Hydrolysis of Proteins

2

Introduction

There are three important principles that apply to *all* limited hydrolytic procedures:

1. Any fragment containing an amino acid residue for which the procedure is specific, that is, one of the "target" amino acids, will have that amino acid at either the amino- or carboxyl-terminal position, depending on the cleavage specificity of the procedure. If the reagent cleaves on the carboxyl side, the target amino acid will be at the *C*-terminus of the product fragment, whereas cleavage on the amino side will yield products with the target amino acid at the *N*-terminus.
2. Any fragment lacking the target amino acid(s) must have arisen from either the *N*- or *C*-terminal region of the original peptide, again depending on the specificity of the reaction. If the reagent is carboxyl specific, fragments lacking the target amino acids will be generated by the original *C*-terminal region; if the reagent is amino specific, fragments lacking the target amino acids will come from the original *N*-terminal region.
3. Any fragment consisting of *only* the target amino acid must have originated as the *N*-terminal amino acid (if cleavage is at the carboxyl side), as the *C*-terminal amino acid (for cleavage at the amino side), or from a region of two or more consecutive target amino acids.

Cyanogen Bromide

Several procedures are available for limited hydrolysis of a peptide chain. Most of these are enzymatic methods of varying specificities, but one important, highly specific chemical method is also widely used, namely treatment with cyanogen bromide (CNBr).

CNBr reacts specifically on the carboxyl side of a methionine-containing peptide, cleaving the chain between methionine and the next amino acid and converting methionine to homoserine lactone. As is usually the case, individual fragments are isolated, and each is subjected to complete hydrolysis in preparation for amino acid analysis. At the same time that peptide bonds are broken, hydrolysis cleaves the lactone to homoserine, which can be detected and quantitated by the amino acid analyzer. In brief, CNBr cleaves peptides on the carboxyl side of methionine, converting methionine to homoserine.

Let us examine the three principles of limited hydrolysis, since an understanding of them is crucial to developing a logical thought process for solving sequencing problems. The current discussion will center on cyanogen bromide but analogous arguments will be made later in connection with other procedures.

Consider peptide A:

Val–Ala–Met–Ala–Val

Treatment with cyanogen bromide will yield two products, namely Val–Ala–Hsr and Ala–Val. (Note that the abbreviation Hsr is used to indicate either homoserine itself or the lactone.)

Principles 1 and 2 are illustrated by this cleavage. The first product, a tripeptide, contains the target amino acid homoserine (which originated, of course, as methionine) at its *C*-terminus (Principle 1). The second product, the dipeptide, was originally at the *C*-terminus of the pentapeptide. Since this fragment was beyond the point of cleavage, it appears as a product lacking the target amino acid (Principle 2).

Consider also the following cyanogen bromide cleavages.

Peptide B:

Gly–Phe–Met–Ala–Tyr–Ser–Met–Thr–Lys → Gly–Phe–Hsr +
Ala–Tyr–Ser–Hsr + Thr–Lys

In both homoserine-containing fragments, the target amino acid is at the *C*-terminus, whereas the fragment arising from the original *C*-terminus contains no homoserine.

Peptide C:

Gly–Phe–Ala–Tyr → Gly–Phe–Ala–Tyr

Peptide C contained no methionine, hence no cleavage occurred.

Peptide D:

Gly–Phe–Ala–Tyr–Met → Gly–Phe–Ala–Tyr–Met

Methionine is at the *C*-terminus; therefore, no cleavage occurs and methionine (rather than homoserine) appears in the amino acid analysis.

Peptide E:

Met–Gly–Phe–Ala–Tyr → Hsr + Gly–Phe–Ala–Tyr

Methionine at the *N*-terminus appears as free homoserine rather than as a homoserine-containing peptide.

Peptide F:

Ser–Phe–Met–Met–Tyr–Gly → Ser–Phe–Hsr + Hsr + Tyr–Gly

Free homoserine arises from the sequence Met–Met (Principle 3).

Example 2.1

Cyanogen bromide treatment of a peptide with the amino acid composition (Ala_2,Met) gives two products. One product contains only alanine, while the other contains alanine and homoserine. What is the sequence of the original peptide?

The product containing only alanine must come from the original *C*-terminus, giving the partial sequence (Met,Ala)–Ala.

The dipeptide product contains homoserine, which must be at its *C*-terminus. The sequence of the dipeptide is therefore Ala–Hsr, and the sequence of the original peptide is Ala–Met–Ala.

Example 2.2

Cyanogen bromide treatment of a peptide with the amino acid composition (Ala_2,Met) gives two products. One product contains only homoserine, while the other contains only alanine. What is the sequence of the original peptide?

The appearance of homoserine as a free amino acid implies *N*-terminal methionine (see peptide E, above). The sequence of the original peptide is Met–Ala–Ala, and the alanine-containing fragment must be the dipeptide Ala–Ala. Since the amino acid analysis shows only alanine, it is not necessary to note that there are really two alanine residues. Data of this type may be misleading; be sure to consider the possibility that a product containing only a single amino acid may, in fact, represent more than one equivalent of that amino acid.

Example 2.3

Cyanogen bromide treatment of a peptide with the amino acid composition (Ala_2,Gly_2,Met) gives two products. One product contains only alanine, while the other contains only glycine and homoserine. What is the sequence of the original peptide?

The alanine-containing fragment must come from the original *C*-terminus, since it contains no homoserine. It is safe to assume that, since the other fragment contains no alanine, the *C*-terminal fragment actually contains two alanine residues and that the *C*-terminal sequence is Ala–Ala (see Example 2.2, above). Homoserine must be at the *C*-terminus of the other (i.e., the *N*-terminal) fragment. Since the *C*-terminal fragment contains no glycine, the *N*-terminal fragment has the composition (Gly_2,Hsr), and its sequence is Gly–Gly–Hsr. The original sequence then is Gly–Gly–Met–Ala–Ala.

Although it is generally not a good idea to make any assumptions about the data presented, this particular problem would be otherwise insoluble. The key to deciding whether a given fragment may be "hiding" something is to compare the original amino acid composition with what the fragmentation data appear to be saying. If there appears to be a discrepancy, it may be resolved by assuming that a given fragment contains more than one equivalent of an amino acid.

Example 2.4

A peptide has the amino acid composition (Ala_2,Met_2,Ser_2). Cyanogen bromide treatment gives two products whose amino acid compositions are (Ala_2,Hsr) and (Met,Ser_2), respectively. What is the amino acid sequence of the original peptide?

The product containing methionine must come from the original *C*-terminus; the *C*-terminal amino acid must be methionine, and the *C*-terminal sequence is Ser–Ser–Met. The other product has the sequence Ala–Ala–Hsr, and arises from the *N*-terminus. The peptide has the sequence Ala–Ala–Met–Ser–Ser–Met.

Example 2.5

A peptide has the amino acid composition (Ala_2,Met_2,Ser_2). Cyanogen bromide treatment gives three products whose amino acid compositions are (Ser), (Ala,Hsr), and (Ala,Hsr,Ser), respectively. FDNB treatment of the original peptide followed by hydrolysis gives DNP–serine. What is the amino acid sequence of the original peptide?

The *C*-terminus is serine, since the only fragment lacking homoserine contains only serine. Each of the other two fragments has *C*-terminal homoserine. Their original sequences were Ala–Met and (Ala,Ser)–Met, respectively. The cyanogen bromide data do not allow us to determine which of these is the *N*-terminal fragment, nor can we deduce the sequence of the tripeptide fragment.

Treatment with FDNB provides all the missing information. The original *N*-terminus is serine. Of the two fragments in question, only the tripeptide contains serine. Thus the tripeptide fragment is at the *N*-terminus, and its sequence is Ser–Ala–Met. The complete sequence is Ser–Ala–Met–Ala–Met–Ser.

Example 2.6

A peptide has the amino acid composition (Ala_3,Met_2,Ser_2). Treatment of the peptide with FDNB followed by hydrolysis gives DNP–alanine. Lithium borohydride reduction of the peptide followed by hydrolysis gives 2-amino-1-propanol.

Cyanogen bromide treatment of the peptide gives three fragments, CB1, CB2, and CB3, whose amino acid compositions are (Ala,Hsr), (Ala,Ser), and (Ala,Hsr,Ser), respectively. When CB3 is treated with FDNB and hydrolyzed, DNP–serine is produced. What is the amino acid sequence of the peptide?

Notice the method of designating the peptide fragments, identifying them according to their source. Thus, for example, fragments resulting from digestion with the enzyme trypsin are referred to as T1, T2, T3, etc., and fragments arising from digestion with chymotrypsin, another enzyme, may be designated CT1, CT2, etc.

Fragment CB2 is the only one lacking homoserine; therefore, it must be at the *C*-terminus. Lithium borohydride treatment of the original peptide yields 2-amino-1-propanol. The *C*-terminal amino acid is alanine. Thus the *C*-terminal sequence is Ser–Ala, and the entire sequence so far can be shown as (CB1,CB3)–Ser–Ala.

Both CB1 and CB3 have *C*-terminal homoserine. Since CB1 is a dipeptide, its sequence is Ala–Hsr.

FDNB treatment of CB3 gives DNP–serine. The sequence of CB3 can now be deduced since it is a tripeptide whose terminal amino acids are both known: CB3 = Ser–Ala–Hsr.

Treatment of the original peptide with FDNB reveals *N*-terminal alanine. This must be the same alanine found at the *N*-terminus of CB1 (since CB3 has *N*-terminal serine). CB1 is then the *N*-terminal fragment, and the entire sequence of the unknown peptide is Ala–Met–Ser–Ala–Met–Ser–Ala.

Example 2.7

A peptide has the amino acid composition (Ala_2,Leu_2,Met_3,Val_3). Treatment of the peptide with FDNB followed by hydrolysis gives DNP–valine. Lithium borohydride reduction of the peptide, followed by hydrolysis, gives 2-amino-3-methyl-1-butanol.

Cyanogen bromide treatment of the peptide gives four fragments whose amino acid compositions are:

CB1 = Hsr
CB2 = (Ala,Hsr,Val)
CB3 = (Hsr,Leu,Val)
CB4 = (Ala,Leu,Val)

FDNB treatment of CB2, CB3, and CB4 followed by hydrolysis yields DNP–alanine, DNP–valine, and DNP–leucine, respectively. Give as much of the amino acid sequence of the original peptide as possible.

From the end-group data we conclude that the *N*- and *C*-termini are both valine.

CB1 contains only homoserine, which could have arisen from either *N*-terminal methionine or from a sequence of two adjacent methionine residues. Since the *N*-terminus is known to be valine, the original peptide must contain a …Met–Met… sequence.

CB4 contains no homoserine; it must be the *C*-terminal fragment. From the FDNB data we conclude that its *N*-terminus is leucine. Since its *C*-terminus is already known (from the lithium borohydride data) to be valine, the sequence of CB4 is Leu–Ala–Val.

The sequences of CB2 and CB3 are readily established from the specificity of cyanogen bromide cleavage (both have *C*-terminal homoserine) and from the subsequent FDNB data, which establish their *N*-termini as alanine and valine, respectively. Thus, the sequence of CB2 is Ala–Val–Hsr and of CB3 is Val–Leu–Hsr. Of these, only CB3 has *N*-terminal valine, and therefore qualifies as the *N*-terminal fragment.

We can write the partial sequence CB3–(CB1,CB2)–CB4. The data do not allow us to deduce the location of CB1 relative to CB2. Thus, the sequence of the original peptide is either Val–Leu–Met–Met–Ala–Val–Met–Leu–Ala–Val or Val–Leu–Met–Ala–Val–Met–Met–Leu–Ala–Val.

Trypsin

Trypsin-catalyzed hydrolysis is the most highly specific of the four enzymatic methods we will discuss. Trypsin catalyzes hydrolysis of the peptide bond on the carboxyl side of either lysine or arginine (i.e., the basic amino acids).

Example 2.8

Tryptic digestion of a peptide with the amino acid composition (Ser_2,Lys) gives two products. One product contains only serine, while the other contains serine and lysine. What is the sequence of the original peptide?

The product containing only serine must come from the original *C*-terminus, giving the partial sequence (Lys,Ser)–Ser. The dipeptide product contains lysine, which must be at its *C*-terminus. The sequence of the dipeptide is then Ser–Lys, and the sequence of the original peptide is Ser–Lys–Ser.

Note that this example is analogous to Example 2.1. In fact, each of the examples shown for cyanogen bromide cleavage could be rewritten in terms of tryptic digestion, generating a whole set of "new" examples. This will not be carried to extremes, however, since trypsin provides the possibility of much more complex and interesting examples.

Example 2.9

A heptapeptide has the amino acid composition (Ala,Arg,Gly, Ile_2,Lys,Val). Reaction with lithium borohydride followed by hydrolysis yields 2-amino-3-methyl-1-pentanol. Treatment with FDNB followed by hydrolysis gives DNP-glycine.

Tryptic digestion gives three fragments, T1, T2, and T3, with amino acid compositions (Ile,Val), (Ala,Arg), and (Gly,Ile,Lys), respectively. What is the amino acid sequence of the heptapeptide?

T1 is the *C*-terminal peptide, since it contains no basic amino acids. The lithium borohydride data indicate *C*-terminal isoleucine so that the sequence of T1 is Val–Ile.

The FDNB experiment shows that the *N*-terminus of the heptapeptide is glycine. Since T3 is the only fragment containing glycine, T3 must be at the *N*-terminus of the heptapeptide. The sequence can be shown as T3–T2–T1.

Both T2 and T3 contain basic amino acids at their *C*-termini (from the specificity of tryptic digestion). This establishes the sequences of T2 as Ala–Arg and T3 as Gly–Ile–Lys. The sequence of the heptapeptide is Gly–Ile–Lys–Ala–Arg-Val–Ile.

Example 2.10

A decapeptide has the amino acid composition (Arg,Glu,Gly, Lys_2,Pro,Ser_2,Thr,Val). Reaction with lithium borohydride followed by hydrolysis yields 2-amino-1,5-pentanediol. Treatment with FDNB followed by hydrolysis gives DNP-serine.

Tryptic digestion gives four fragments, T1, T2, T3, and T4, with amino acid compositions (Glu,Ser), (Lys,Val), (Arg,Gly,Ser), and (Lys,Pro,Thr), respectively. How much of the amino acid sequence of the decapeptide can be deduced?

Fragment T1, which lacks basic amino acids, is at the *C*-terminus. The lithium borohydride data indicate *C*-terminal glutamate (both for the original peptide and for T1), so that the sequence of T1 is Ser–Glu.

Fragment T2 is a dipeptide. Since lysine is at its *C*-terminus, its sequence must be Val–Lys.

The FDNB data show that the *N*-terminus of the original peptide is serine. Both T1 and T3 contain serine, but T1 has already been established as the *C*-terminal fragment. Thus T3 is Ser–Gly–Arg and is the *N*-terminus.

Fragment T4 has lysine at its *C*-terminus, but the data do not allow assignment of the other two amino acid residues. Thus a partial sequence for T4 is (Pro,Thr)–Lys. It is also not known whether the overall sequence is

T3–T4–T2–T1 = Ser–Gly–Arg–(Pro,Thr)–Lys–Val–Lys–Ser–Glu

or

T3–T2–T4–T1 = Ser–Gly–Arg–Val–Lys–(Pro,Thr)–Lys–Ser–Glu

Additional data are needed to solve this problem, but notice how this example undergoes a radical change if methionine replaces threonine (Example 2.11).

Example 2.11

A decapeptide has the amino acid composition (Arg,Glu,Gly, Lys_2,Met,Pro,Ser_2,Val). Reaction with lithium borohydride followed by hydrolysis yields 2-amino-1,5-pentanediol. Treatment with FDNB followed by hydrolysis gives DNP–serine.

Tryptic digestion gives four fragments, T1, T2, T3, and T4, with amino acid compositions (Glu,Ser), (Lys,Val), (Arg,Gly,Ser), and (Lys,Met,Pro), respectively. Treatment of the decapeptide with CNBr gives two fragments, CB1 and CB2, with amino acid compositions (Glu,Lys,Pro,Ser) and (Arg,Gly,Hsr,Lys,Ser,Val), respectively. What is the amino acid sequence of the decapeptide?

With the exception of the replacement of threonine by methionine, this peptide is identical to the one in the preceding example. The trypsin digestion data are identical (except, again, for the single replacement), so that, excluding the CNBr data, we are left with the two possibilities given above, namely

T3–T4–T2–T1 = Ser–Gly–Arg–(Met,Pro)–Lys–Val–Lys–Ser–Glu

or

T3–T2–T4–T1 = Ser–Gly–Arg–Val–Lys–(Met,Pro)–Lys–Ser–Glu

Now examine the cyanogen bromide data. Fragment CB1 contains only four amino acids. If the first possible sequence were correct, then methionine would be in either position 4 or position 5. One of the products of the CNBr reaction would then be either a tetrapeptide containing homoserine (if methionine were at position 4) or a pentapeptide containing homoserine (if methionine were at position 5). The actual product, though, is a hexapeptide containing homoserine. Since neither of these proposed sequences accounts for the products, neither can be correct.

The second possibility is

T3–T2–T4–T1 = Ser–Gly–Arg–Val–Lys–(Pro,Met)–Lys–Ser–Glu

which has methionine in either position 6, which would give a hexapeptide containing homoserine, or position 7, which would give a heptapeptide containing homoserine. The first of these corresponds to the actual experimental results. The sequence of the decapeptide is Ser–Gly–Arg–Val–Lys–Met–Pro–Lys–Ser–Glu.

The presence of methionine allows a second cleavage reaction to complement the data obtained from tryptic digestion and provides sufficient information to solve the entire sequence. In general, more complex peptides require at least two, and often more, cleavage reactions.

This procedure, i.e., use of two or more complementary cleavages, is generally referred to as the "overlap" method, since the results of one experiment provide data that can overlap, and clarify, the results of another.

Analysis of the data need not follow the order in which they are presented; in many cases it is advantageous to consider the data in some other order. Suppose, in the previous example, we consider the cyanogen bromide experiment first.

The two cyanogen bromide fragments are readily ordered since CB1 contains no homoserine and must therefore be the *C*-terminal fragment. The specificity of the reaction immediately indicates that homoserine must be at the *C*-terminus of CB2. A partial sequence is

CB2–CB1 = (Arg,Gly,Lys,Ser,Val)–Hsr–(Glu,Lys,Pro,Ser)

Adding the end-group data:

CB2–CB1 = Ser–(Arg,Gly,Lys,Val)–Met–(Lys,Pro,Ser)–Glu

The tryptic digestion data are now much more clear:

T1 = Ser–Glu (*C*-terminal)
T3 = Ser–Gly–Arg (*N*-terminal)
T2 = Val–Lys (between T3 and T4)
T4 = Met–Pro–Lys

Example 2.12

The amino acid composition of a peptide is (Ala_2,Arg,Asp, Glu_2,Gly_2,Lys_2,Met_2,Phe). Reduction of the peptide with lithium borohydride followed by hydrolysis gives three new products,

Table 1

Lithium Borohydride Reduction Products of the Acidic Amino Acids and their Amides

Asp (internal)	→	2-amino-4-hydroxybutanoic acid
Asp (*C*-terminal)	→	2-amino-1,4-butanediol
Asn (internal)	→	aspartic acid
Asn (*C*-terminal)	→	3-amino-4-hydroxybutanoic acid
Glu (internal)	→	2-amino-5-hydroxypentanoic acid
Glu (*C*-terminal)	→	2-amino-1,5-pentanediol
Gln (internal)	→	glutamic acid
Gln (*C*-terminal)	→	4-amino-5-hydroxypentanoic acid

namely 2-amino-4-hydroxybutanoic acid, 2-amino-5-hydroxy-pentanoic acid, and 4-amino-5-hydroxypentanoic acid.

Cyanogen bromide cleavage yields three fragments with the following amino acid compositions:

CB1 = (Glu,Gly,Lys)
CB2 = (Ala,Glu,Hsr,Lys,Phe)
CB3 = (Ala,Arg,Asp,Gly,Hsr)

Digestion with trypsin gives four fragments. Hydrolysis of each followed by amino acid analysis gives the following data:

T1 = Glu
T2 = (Glu,Lys)
T3 = (Asp,Gly_2,Lys,Met)
T4 = (Ala_2,Arg,Met,Phe)

Treatment of T3 with FDNB followed by hydrolysis yields DNP–glycine, while similar treatment of T4 gives DNP–alanine. Give the complete sequence of the peptide.

Of the three lithium borohydride reduction products, the only one of any significance is the third, namely 4-amino-5-hydroxypentanoic acid. The other two products (2-amino-4-hydroxybutanoic acid and 2-amino-5-hydroxypentanoic acid) arose from internal aspartate and glutamate residues, respectively (see Table 1). But 4-amino-5-hydroxypentanoic acid originated as *C*-terminal glutamine. Thus, while the amino acid analysis of the original peptide shows two glutamate residues, one of these corresponds to the *C*-terminal glutamine.

We can deduce a partial sequence for fragment CB1. Since homoserine is lacking, CB1 must have been the original *C*-terminal fragment. Therefore the Glu in CB1 must be the *C*-terminal glutamine: CB1 = (Gly,Lys)–Gln.

CB2 and CB3, of course, contain *C*-terminal homoserine, so that their sequences would be (Ala,Glu,Lys,Phe)–Met and (Ala,Arg,Asp,Gly)–Met, respectively. Two possible partial sequences may be written for the original peptide:

CB2–CB3–CB1 = (Ala,Glu,Lys,Phe)–Met–(Ala,Arg,Asp,Gly)–Met–(Gly,Lys)–Gln

or

CB3–CB2–CB1 = (Ala,Arg,Asp,Gly)–Met–(Ala,Glu,Lys,Phe)–Met–(Gly,Lys)–Gln

No further information can be extracted from the CNBr data.

Fragment T1 must be the *C*-terminus since it contains no basic amino acids. From the lithium borohydride data on the original peptide, T1 is Gln. Fragments T2, T3, and T4 must all contain basic amino acids at their *C*-termini. Thus,

T2 = Glu–Lys
T3 = (Asp,Gly_2,Met)–Lys
T4 = (Ala_2,Met,Phe)–Arg

The only arginine residue in the peptide is at the *C*-terminus of the pentapeptide T4. If T4 were at the original *N*-terminus, arginine would be no closer to the *N*-terminus than position 5. Of course, if T4 is not the *N*-terminal fragment, the arginine will be even further from the *N*-terminus.

In fragment CB3 arginine is at position 1, 2, 3, or 4. However, since we have seen that arginine must be no closer to the *N*-terminus than position 5, CB3 cannot be at the *N*-terminus, and we can eliminate the sequence CB3-CB2-CB1 from consideration. The only remaining possibility is CB2–CB3–CB1.

The *C*-terminal sequence of the original peptide is ...Met–(Gly,Lys)–Gln. The glutamine at the *C*-terminus has already been identified as fragment T1. Of the remaining tryptic fragments, only T3 contains methionine, glycine, and lysine. Thus T3 must directly precede T1. Furthermore, the lysine at the *C*-terminus of T3 must come "after" (i.e., on the carboxyl side of) both methionine and glycine. Since methionine precedes glycine, the *C*-terminal sequence must be ...Met–Gly–Lys–Gln, and the partial sequence T3–T1 is (Asp,Gly)–Met–Gly–Lys–Gln.

FDNB data show that the *N*-terminus of T3 is glycine; thus T3-T1 = Gly–Asp–Met–Gly–Lys–Gln, and the partial sequence of the original peptide can be shown as

CB2–CB3–CB1 = (T2,T4)–T3–T1 =
(Ala,Glu,Lys,Phe)–Met–(Ala,Arg)–Gly–Asp–Met–Gly–Lys–Gln

Fragment T2 contains the sequence Glu–Lys. It is apparent that these two amino acids must precede the methionine at the *C*-terminus of CB2 (i.e., the methionine at position 5). The methionine residue at position 5 is also contained in T4. Since the dipeptide Glu–Lys of T2 must precede the methionine of T4, we conclude that fragment T2 precedes fragment T4. The partial sequence can be rewritten:

T2–T4–T3–T1 = Glu–Lys–(Ala,Phe)–Met–(Ala,Arg)–Gly–Asp–Met–Gly–Lys–Gln

The only remaining uncertainties are in fragment T4. Since this is a tryptic fragment, arginine must be at the *C*-terminus, and alanine must therefore be between methionine and arginine. The sequence can now be shown as T4 = (Ala,Phe)–Met–Ala–Arg. The FDNB experiment shows the *N*-terminus of T4 to be alanine. The sequence of T4 is Ala–Phe–Met–Ala–Arg and the sequence of the entire peptide is Glu–Lys–Ala–Phe–Met–Ala–Arg–Gly–Asp–Met–Gly–Lys–Gln.

Note the way that this analysis eliminated one of the two possible sequences arising from the cyanogen bromide experiment. Once the composition and partial sequence of fragment T4 became known, it was possible to show that the presence of arginine at the *C*-terminus of T4 was consistent with only one of the two possible CNBr sequences. The remaining possibility had to be discarded. Arguments of this type are very important and will be utilized again. Note also the fact that lithium borohydride data can be confusing. It is generally a good idea to show the structures of the reduction products in order to determine their sources.

Chymotrypsin

The action of chymotrypsin is analogous to that of trypsin, but the specificity is slightly broader. Whereas trypsin is specific for the two basic amino acids, lysine and arginine, chymotrypsin catalyzes the hydrolysis of peptide bonds in which the carboxyl group is contributed by one of the three aromatic amino acids, namely phenylalanine, tyrosine, or tryptophan.

Example 2.13

Chymotryptic digestion of a peptide with the amino acid composition (Ile_2,Phe) gives two products. One product contains only isoleucine, while the other contains isoleucine and phenylalanine. What is the sequence of the peptide?

The product containing only isoleucine must come from the original *C*-terminus, giving the partial sequence (Ile,Phe)–Ile. The dipeptide product contains phenylalanine, which must be at its *C*-terminus. Therefore, the sequence of the dipeptide is Ile–Phe, and the sequence of the original peptide is Ile–Phe–Ile.

This problem is analogous to those presented in Examples 2.1 and 2.8. It should be apparent by now that a problem of this type is trivial.

Example 2.14

Chymotryptic digestion of a pentapeptide with the amino acid composition (Ala_2,Tyr,Val_2) gives two fragments. One fragment

contains equimolar amounts of alanine and tyrosine. Treatment of the other fragment with FDNB gives DNP–valine. Lithium borohydride reduction of the original peptide gives 2-amino-3-methyl-1-butanol. What is the sequence of the pentapeptide?

Since the original pentapeptide contains only one tyrosine residue, the fragment containing "equimolar amounts of alanine and tyrosine" must be either the dipeptide Ala–Tyr, the tripeptide (Ala,Val)–Tyr, or the tetrapeptide (Ala,Val_2)–Tyr. The last possibility is eliminated upon finding that at least one valine is present in the second product. In either of the two remaining possibilities tyrosine is at the *C*-terminus. Note that no more than one residue of alanine may appear in this fragment.

The "other fragment" contains no tyrosine and, therefore, originated at the *C*-terminus of the pentapeptide. Its amino acid composition is either (Ala,Val) or (Ala,Val_2). Its *N*-terminus, from the FDNB experiment, is valine, as is its *C*-terminus (from the lithium borohydride experiment). This fragment thus contains both valine residues, as well as one alanine residue. Its sequence is Val–Ala–Val, and the sequence of the pentapeptide is Ala–Tyr–Val–Ala–Val.

Example 2.15

A decapeptide has the amino acid composition (Ala_2,Gly_2, Met,Phe,Ser,Tyr,Val_2). The decapeptide is treated with cyanogen bromide, and an aliquot of the product mixture is treated with FDNB. After hydrolysis the only yellow product to appear is DNP–glycine. The remainder of the cyanogen bromide product mixture is resolved in the usual way, yielding two fragments whose amino acid compositions are:

CB1 = (Ala,Gly,Tyr,Val)
CB2 = (Ala,Gly,Hsr,Phe,Ser,Val)

Digestion of the decapeptide with chymotrypsin gives three fragments. Their amino acid compositions are:

CT1 = Ala
CT2 = (Gly,Phe,Ser)
CT3 = (Ala,Gly,Met,Val_2,Tyr)

After treatment of CT3 with FDNB, hydrolysis gives DNP–alanine. What is the amino acid sequence of the pentapeptide?

There is no reason in the world to treat an unresolved fragmentation mixture with any end-group reagent. Nonetheless this makes an interesting way of providing data for a hypothetical problem. The fact that the mixture of CNBr fragments gives only a single DNP-amino acid indicates that all fragments in that mixture contain the same *N*-terminal amino acid. From subsequent data we find that there are two CNBr fragments, so we conclude that each contains *N*-terminal glycine. CB1, which does not contain homoserine, is the *C*-terminal fragment, and the partial sequence is

CB2–CB1 = Gly–(Ala,Phe,Ser,Val)–Met–Gly–(Ala,Tyr,Val)

Fragment CT1, which contains no aromatic amino acids, is at the original *C*-terminus:

CB2–CB1 = Gly–(Ala,Phe,Ser,Val)–Met–Gly–(Tyr,Val)–Ala

The sequence of CT3 is (Ala,Val)–Met–Gly–Val–Tyr. Since tyrosine must be at the *C*-terminus of CT3, the valine residue in the sequence (Tyr,Val) must lie between glycine and tyrosine (i.e., at position 5 of CT3).

The FDNB experiment shows that the *N*-terminus of CT3 is alanine, so the sequence of this fragment is Ala–Val–Met–Gly–Val–Tyr.

Fragment CT2 is at the *N*-terminus of the decapeptide. Since an *N*-terminal glycine has already been established, the sequence of CT2 is Gly–Ser–Phe, and the sequence of the decapeptide is

CT2–CT3–CT1 = Gly–Ser–Phe–Ala–Val–Met–Gly–Val–Tyr–Ala

Elastase

Elastase functions in a manner analogous to trypsin and chymotrypsin, cleaving on the carboxyl (i.e., "right") side of its target amino acids. The enzyme is specific for amino acids containing small hydrophobic side chain groups (alanine, valine, and glycine). As in the case of cyanogen bromide, trypsin, and chymotrypsin, the products contain the target amino acids at their *C*-termini.

Pepsin

The specificity of pepsin is both broader than and qualitatively different from that of any of the reagents considered so far. Pepsin will digest proteins by catalyzing hydrolysis of peptide bonds involving leucine, the aromatic amino acids (phenylalanine, tyrosine, or tryptophan), or the acidic amino acids (aspartate or glutamate), but not their amides (asparagine or glutamine). The fact that pepsin cleaves at any of six amino acids means that, in general, cleavage with pepsin will yield more fragments than with any other reagent. This may or may not be advantageous since a greater quantity of fragments may lead to more confusion than illumination. On the other hand, peptic digestion may often give information that cannot be obtained by other methods.

In addition to having a broader range of susceptible amino acids, pepsin also cleaves in a manner different from those of the other reagents discussed so far.

Whereas cyanogen bromide, trypsin, chymotrypsin, and elastase all cleave to the carboxyl (i.e., "right") side of their target amino acids, pepsin cleaves to the amino (i.e., "left") side of its target amino acids. As a result the target amino acids will be at the *N*-termini of the peptic fragments.

Example 2.16

Peptic digestion of a peptide with the amino acid composition (Ile_2,Phe) gives two products. One product contains only isoleucine, while the other contains isoleucine and phenylalanine. What is the sequence of the peptide?

The product containing only isoleucine must come from the original *N*-terminus, giving the partial sequence Ile–(Phe,Ile). The dipeptide product contains phenylalanine, which must be at its *N*-terminus. The sequence of the dipeptide is then Phe–Ile, and the sequence of the original peptide is Ile–Phe–Ile.

Example 2.17

A peptide with the amino acid composition (Glu,Ile,Leu, Phe,Ser,Val) is digested with pepsin, giving three fragments. Their amino acid compositions are:

P1 = Ile
P2 = (Glu,Leu)
P3 = (Phe,Ser,Val)

Lithium borohydride reduction of the original peptide followed by hydrolysis gives 2-amino-1,3-propanediol. What is the amino acid sequence of the peptide?

Fragment P2 stands out as a apparent paradox in that it contains two of pepsin's target amino acids. What this means, of course, is that glutamic acid is merely an artifact of the hydrolysis/amino acid analysis procedure, and it is not really one of the amino acids in the original peptide. Glutamic acid must arise by hydrolysis of its amide, glutamine. Since pepsin does not recognize either glutamine or asparagine, appearance of either glutamate or aspartate in a peptic fragment that also contains one of the other target amino acids yields an apparently anomalous amino acid composition. The composition of P2 can be restated as (Gln,Leu). Since leucine must be at the *N*-terminus of P2, the sequence is Leu–Gln.

Lithium borohydride reduction of the original peptide reveals the *C*-terminus to be serine. This also defines the sequence of fragment P3: P3 = Phe–Val–Ser.

Fragment P1, containing only isoleucine, must be at the original *N*-terminus so that the entire sequence is Ile–Leu–Gln–Phe–Val–Ser.

Notice that if chymotrypsin were used to digest this peptide only two fragments would be observed, a dipeptide (Val–Ser) and a tetrapeptide (Ile–Leu–Gln–Phe), and that the sequence of the tetrapeptide could be deduced only from additional experimental data such as a sequential Edman degradation.

End of Chapter Problems

Problem 2.1

In each of the following experiments the peptide

Pro–Gly–Thr–Lys–Asn–Met–Val–Phe–Pro–Glu–Gly–Leu–Lys

is treated with the reagent listed. All fragments are isolated, hydrolyzed, and subjected to amino acid analysis. Predict the results of the following analyses:

A. Cyanogen bromide
B. Trypsin
C. Chymotrypsin
D. Elastase
E. Pepsin

Problem 2.2

A heptapeptide has the amino acid composition (Ala,Gly,Ile_2,Met_2,Val). Treatment with FDNB gives DNP–glycine. Reduction with lithium borohydride followed by hydrolysis gives 2-amino-3-methyl-1-pentanol. Cyanogen bromide treatment gives three fragments (CB1, CB2, and CB3) with the following amino acid compositions:

CB1 = (Ile,Val)
CB2 = (Ala,Hsr)
CB3 = (Gly,Hsr,Ile)

What is the amino acid sequence of the heptapeptide?

Problem 2.3

A heptapeptide with the amino acid composition (Ala,Arg,Asp,Lys,Pro,Ser,Val) gives three fragments when digested with trypsin. Their compositions are:

T1 = (Asp,Val)
T2 = (Arg,Pro)
T3 = (Ala,Lys,Ser)

Treatment of the heptapeptide with FDNB followed by hydrolysis gives DNP–serine; lithium borohydride reduction of the heptapeptide followed by hydrolysis gives 2-amino-1,4-butanediol. What is the sequence of amino acids in the heptapeptide?

Problem 2.4

The amino acid composition of a nonapeptide is (Arg,Gly_3,Thr,Tyr_2,Ser,Val). Treatment of the nonapeptide with FDNB followed by hydrolysis gives DNP–glycine. Digestion of the nonapeptide with chymotrypsin gives the following three fragments:

CT1 = (Arg,Ser,Tyr)
CT2 = (Gly_2,Thr)
CT3 = (Gly,Tyr,Val)

Treatment of CT1 with FDNB followed by hydrolysis gives DNP–arginine, while treatment of CT2 with FDNB followed by hydrolysis gives DNP–threonine. Deduce the amino acid sequence of the nonapeptide.

Problem 2.5

A decapeptide has the amino acid composition (Ala,Arg_2,Gly,Met_2,Pro,Ser,Tyr_2). Treatment with FDNB followed by hydrolysis gives DNP–serine. Elastase digestion of the decapeptide gives three products:

E1 = Tyr
E2 = (Arg_2,Gly,Met)
E3 = (Ala,Met,Pro,Ser,Tyr)

Treatment of the decapeptide with cyanogen bromide gives three products:

CB1 = (Hsr,Ser,Tyr)
CB2 = (Arg,Gly,Tyr)
CB3 = (Ala,Arg,Hsr,Pro)

What is the amino acid sequence of the decapeptide?

Problem 2.6

A decapeptide has the amino acid composition (Arg,Glu,Gly,Met_3,Ser_2,Thr,Val). Reaction with lithium borohydride followed by hydrolysis gives 2-amino-1,5-pentanediol, while reaction with FDNB gives DNP–serine.

Cyanogen bromide treatment gives four fragments (CB1, CB2, CB3, and CB4) with the following amino acid compositions:

CB1 = (Glu,Ser)
CB2 = (Hsr,Val)
CB3 = (Gly,Hsr,Ser)
CB4 = (Arg,Hsr,Thr)

Digestion with trypsin gives two fragments, whose amino acid compositions are:

T1 = (Arg,Gly,Met,Ser)
T2 = (Glu,Met_2,Ser,Thr,Val)

What is the amino acid sequence of the decapeptide?

Problem 2.7

Peptide X has the amino acid composition (Ala$_2$,Arg$_2$,Asp$_2$,Cys$_4$,Glu,Gly,His$_2$,Ile$_2$,Lys$_2$,Met$_2$,Phe$_2$,Pro$_3$,Ser$_3$,Thr$_3$,Trp,Tyr,Val). Thiol analysis indicates that X contains two disulfide bridges. Treatment of X with mercaptoethanol reduces the disulfides and yields two peptide fragments, Y and Z, which are isolated, hydrolyzed, and subjected to amino acid analysis.

X = (Asp,Cys$_2$,Glu,Gly,His,Ile,Lys$_2$,Phe,Pro$_2$,Ser,Thr$_2$,Trp,Val)
Y = (Ala$_2$,Arg$_2$,Asp,Cys$_2$,His,Ile,Met$_2$,Phe,Pro,Ser$_2$,Thr,Tyr)

Treatment of Y with trypsin gives three fragments, whose amino acid compositions are:

YT1 = (Asp,His,Thr)
YT2 = (Glu,Lys,Pro,Thr)
YT3 = (Cys$_2$,Gly,Ile,Lys,Phe,Pro,Ser,Trp,Val)

Chymotryptic digestion of Y also gives three fragments. Their amino acid compositions are:

YCT1 = (Ile,Trp,Val)
YCT2 = (Cys,Gly,Phe,Pro,Ser)
YCT3 = (Asp,Cys,Glu,His,Lys$_2$,Pro,Thr$_2$)

An elastase digest of Y gives three fragments with the amino acid compositions:

YE1 = (Gly,Ser)
YE2 = (Cys,Phe,Pro,Val)
YE3 = (Asp,Cys,Glu,His,Ile,Lys$_2$,Pro,Thr$_2$,Trp)

Digestion of Y with pepsin gives four fragments with the following amino acid compositions:

YP1 = (Ile,Phe,Val)
YP2 = (Cys,Gly,Pro,Ser)
YP3 = (Cys,Lys,Thr,Trp)
YP4 = (Asp,Glu,His,Lys,Pro,Thr)

Reduction of YP2 with lithium borohydride followed by hydrolysis gives 2-amino-3-mercapto-1-propanol.

In a separate experiment Y is treated with amidase, which catalyzes the hydrolysis of all side chain amides. Digestion of the deamidated peptide with pepsin yields five fragments with the amino acid compositions:

YAP1 = (Asp,His)
YAP2 = (Ile,Phe,Val)
YAP3 = (Cys,Gly,Pro,Ser)

YAP4 = (Cys,Lys,Thr,Trp)
YAP5 = (Glu,Lys,Pro,Thr)

Treatment of fragment Z with cyanogen bromide gives three fragments with the following amino acid compositions:

ZCB1= (Ala,Thr)
ZCB2= (His,Hsr)
ZCB3= (Ala,Arg_2,Asp,Cys_2,Hsr,Ile,Phe,Pro,Ser_2,Tyr)

When Z is digested with chymotrypsin the three fragments obtained have the amino acid compositions:

ZCT1 = (Ala,Ile,Tyr)
ZCT2 = (Ala,Cys,Met,Thr)
ZCT3 = (Arg_2,Asp,Cys,His,Met,Phe,Pro,Ser_2)

Tryptic digestion of Z gives three fragments:

ZT1 = (Arg,Asp,Pro,Ser)
ZT2 = (Arg,His,Met,Ser)
ZT3 = (Ala_2,Cys_2,Ile,Met,Phe,Thr,Tyr)

A peptic digest of Z contains four fragments with the following amino acid compositions:

ZP1 = (Ala,Ile,Phe)
ZP2 = (Arg,Asp,Cys,Ser)
ZP3 = (Ala,Cys,Met,Thr,Tyr)
ZP4 = (Arg,His,Met,Pro,Ser)

Reduction of ZP1 with lithium borohydride followed by hydrolysis gives 2-amino-1-propanol, while similar treatment of ZP3 gives 2-amino-1,3-butanediol.

When unreduced peptide X is treated with chymotrypsin four products are obtained, each of which has a single aromatic amino acid residue.

Give the amino acid sequences of peptide fragments Y and Z and describe the disulfide bonding arrangement of peptide X.

Problem 2.8

Each year I offer my students the opportunity to construct their own peptide sequence problems, both as an intellectual exercise and to convince them that the problems I write actually require some work on my part. This problem, written by Eda Sauterne in 1982, is the most sophisticated one I have ever received. I have assigned it to my classes for several years with the caveat that they should not invest more than two hours in solving it. The problem is presented exactly as Eda submitted it.

Composition of peptide E:

$$(Ala_4,Arg,Asp,Glu,Gly_7,Leu,Lys_2,Phe,Ser_8,Trp,Tyr)$$

Peptic fragments P1 to P7:

P1 = (Ala,Gly,Ser)
P2 = (Ala,Gly,Phe,Ser)
P3 = (Ala,Gly,Ser,Trp)
P4 = (Ala,Gly,Ser,Tyr)
P5 = (Asp,Gly,Lys,Ser)
P6 = (Gly,Leu,Lys,Ser)
P7 = (Arg,Glu,Gly,Ser_2)

When a mixture of fragments P1 to P7 is treated with lithium borohydride, the only product obtained that has no carboxyl group is 2-aminoethanol.
Chymotryptic fragments CT1 to CT4:

CT1 = (Ala,Gly,Ser)
CT2 = (Ala,Asp,Gly_2,Lys,Ser_2,Tyr)
CT3 = (Ala,Gly_2,Leu,Lys,Ser_2,Trp)
CT4 = (Ala,Arg,Glu,Gly_2,Phe,Ser_3)

When a mixture of fragments CT1 to CT4 was treated with FDNB and hydrolyzed the only yellow product obtained is DNP–alanine.

Tryptic fragments T1 to T4:

T1 = (Ala,Gly_2,Ser,Trp)
T2 = (Ala,Arg,Glu,Gly,Ser_2)
T3 = (Ala,Gly_2,Leu,Lys,Ser_2,Tyr)
T4 = (Ala,Asp,Gly_2,Lys,Phe,Ser_3)

What is the amino acid sequence of peptide E?

The two problems that follow represent an attempt on my part to inject a touch of levity (some might say insanity) into biochemistry. The more sophisticated and/or literate students have *appeared* to enjoy these little essays, although I have never been convinced that they were not merely humoring me. In any case, they provide an opportunity to look at sequencing in a somewhat different way, and whether or not they are really "fun," I think they are worthwhile.

Problem 2.9

Recent deep sea exploration of the ill-fated cruise ship *Titanic* resulted in the recovery of the notes of the infamous Bavarian biochemist Heinrich Dasgesass. Dasgesass was known in his day as a rather eccentric protein chemist whose ideas were either totally absurd or incredibly foresighted. His supporters called him brilliant, while his detractors labeled him a quack. He claimed, for example, in 1906 to have invented a crude but effective amino acid analyzer which was powered by a bizarre assembly involving laboratory rats on a treadmill.

Although the recovered research notes were partially obliterated by decades of immersion in sea water, it was possible to reconstruct one section of Dasgesass's data. The legible text is reproduced here.

An unusual pentapeptide has been observed in an extract of the rare plant *an&#b?s s%t*v#. Complete hydrolysis and amino acid analysis of this peptide were not sufficient to determine the sequence of amino acids. Thus it became necessary to treat the peptide with the endopeptidase g&d*~k>j%. This treatment yielded two products, which were separated, individually hydrolyzed, and subjected to amino acid analysis. Lo and behold, after this single cleavage step it became clear that the sequence of amino acids in the peptide was...

Here the account becomes totally illegible. What would you suggest as a possible sequence for this peptide? Justify your answer on the basis of Dasgesass's experimental data and your knowledge of peptide sequencing chemistry.

Problem 2.10

Dr. Crusoe Robinson, the famous biochemist, was washed ashore with his travel kit containing trypsin, chymotrypsin, and pepsin, a few appropriate buffers, some glassware, a short but highly efficient gel filtration column, and his trusty amino acid analyzer, all of which he managed to salvage before his sea-going protein sequencing laboratory sank in a typhoon. Shortly after he set up shop on his new island home, the chieftain of the local native tribe appeared at Robinson's door, carrying a small vial containing a sample of a peptide. The tribe's witch doctor, who had isolated the peptide, claimed to have seen its sequence in a hallucinogen-induced dream during a recent orgy. The peptide sequence he saw was Cys–Tyr–Phe–Gln–Asp–Lys–Pro–Arg–Gly.

The chieftain, skeptical of the witch doctor's scientific abilities, had come to Robinson for verification of the sequence. Robinson, for his part, was fond of the witch doctor, who was, after all, the closest thing to a professional colleague he had on the island.

"Bring me the true sequence with all the appropriate experimental data to justify it within three days," uttered the chieftain ominously.

Robinson, whose many other obligations included coaching the tribe's water polo and chess teams, was unable to present his results to the chieftain at the appointed time. He sent his assistants, the Siamese-twin Day sisters, Summe and Ennie, to convey his respects and show his research notes to the chieftain.

"Milord," began Ennie, "Dr. Robinson is happy to verify the witch doctor's sequence. We needed only three enzyme-catalyzed cleavage experiments to determine the entire sequence."

"After each cleavage," continued Summe, "we separated the products by gel filtration, then hydrolyzed each one and determined its amino acid composition."

Briefly summarize Robinson's experiments. For each enzyme-catalyzed reaction give the amino acid composition found for each hydrolyzed cleavage product. Explain how the data supported the witch doctor's proposed sequence.

Base Sequencing of DNA

Introduction

The determination of the sequence of bases in a strand of DNA has developed to the point at which it is an almost trivial operation. Although rather deft manipulative techniques are required for success, the theory of the procedure is rather straightforward; however, it requires a type of thinking that is, at first, alien to students of biochemistry. Throughout all your chemistry and biology courses you are trained to think of all molecules of a given substance as reacting in exactly the same way. Random variations are generally considered minor irritants at best. But in the sequencing of DNA it is crucial that we consider each molecule in a sample as an independent entity which can react in a manner distinct from that of every other molecule. Keep this in mind as you read the theory.

Two sequencing methods were developed early on. While their philosophies and approaches differ, the results they present in the laboratory have quite similar appearances. Other methods, requiring high-powered instrumentation, have recently become available. But, since their object is to give unambiguous results requiring little or no interpretation or understanding on the part of the experimenter, we will not discuss them.

Both "classical" methods, the one developed by Maxam and Gilbert and the other developed by Sanger, rely on electrophoresis of the product mixtures on polyacrylamide gels which are capable of resolving oligonucleotides differing in size by as little as a single nucleotide. In either procedure four reaction mixtures are prepared, each containing a collection of radioactively labeled oligonucleotides. Application to each of four wells of a polyacrylamide gel is followed by electrophoresis and autoradiography, resulting in a permanent record of the gel.

Electrophoresis is carried out at alkaline pH in the presence of the denaturing detergent sodium dodecyl sulfate (SDS). As a result, all oligonucleotide fragments are negatively charged, and their migration rates are determined by their sizes so that smaller fragments, which encounter less resistance, move more rapidly through the gel than larger ones. Since the product mixture in each of the four reaction samples is specific for a particular base or combination of bases, it is possible to read the sequence of bases directly from the autoradiogram, starting with the 5′-end at the bottom of the gel, and working back to the 3′-end. Some specific examples will make this much easier to understand.

The Maxam–Gilbert Method

In the Maxam–Gilbert (chemical) method, the DNA sample is first dephosphorylated at both 5′-ends by the action of alkaline phosphatase, and is then rephosphorylated with radioactive phosphate by treating with [^{32}P]-ATP in the presence of polynucleotide kinase. The sample is denatured and the two labeled strands are separated from each other by electrophoresis. The resolved strands are extracted from the gel, divided into four aliquots each, and treated with modifying reagents under conditions chosen so that each molecule reacts at only a single, random site.

Dimethyl sulfate attacks guanine; formic acid reacts with both adenine and guanine; hydrazine in the presence of sodium chloride is specific for cytosine; and hydrazine alone reacts with both cytosine and thymine. In each case the primary reaction is followed by treatment with piperidine, which displaces the modified bases and quantitatively cleaves their phosphodiester bonds so that the fragment retains the 3′-phosphate originally associated with the displaced base. In each of the four reaction tubes representing each strand, these treatments yield a mixture of all possible products in which cleavage at a particular base can occur. For example, if the sequence of one strand of the original DNA were pTpCpTpApCpGpCpTpGpT then the product mixtures would be as follows (the * represents the [^{32}P] label):

Dimethyl sulfate/piperidine product (G lane)	*pTpCpTpApCp(Gp)
	*pTpCpTpApCpGpCpTp(Gp)
Formic acid/piperidine products (A + G lane)	*pTpCpTp(Ap)
	*pTpCpTpApCp(Gp)
	*pTpCpTpApCpGpCpTp(Gp)
Hydrazine/piperidine products (C + T lane)	*p(Tp)
	*pTp(Cp)
	*pTpCp(Tp)
	*pTpCpTpAp(Cp)
	*pTpCpTpApCpGp(Cp)
	*pTpCpTpApCpGpCp(Tp)
	*pTpCpTpApCpGpCpTpGp(T)
Hydrazine/NaCl/piperidine products (C lane)	*pTp(Cp)
	*pTpCpTpAp(Cp)
	*pTpCpTpApCpGp(Cp)

There are three important points to recognize in these product mixtures:

1. The base shown in parentheses at the 3′-end of each product is the one which has been removed in the cleavage process. What remains is just the sugar phosphate residue originating at that nucleotide. Knowing the specificity of the reaction, however, makes it possible to deduce the base that had been present.
2. Every possible oligonucleotide product, from 1 to 10 nucleotides in length, is found in the combined set of products. Some products are found in only one mixture each, while others are found in a pair of mixtures. For example, the pentanucleotide *pTpCpTpAp(Cp) is found in both the hydrazine/piperidine product mixture and in the hydrazine/NaCl/piperidine product mixture. Thus, upon electrophoresis, the lanes representing both reagent mixtures will show a band at the same position.
3. The sequence that will be read from the autoradiogram is that of the unknown DNA strand, not, as is the case in the Sanger method, that of its antiparallel complement.

The four product mixtures are then subjected to electrophoresis on a polyacrylamide gel, after which the gel is placed on a sheet of X-ray film for autoradiography. Development of the film yields a set of bands representing the oligonucleotide products. A band appearing in both the G and A + G lanes indicates a fragment having G at its 3′-end, while a band in only the A + G lane signifies 3′-A. Similarly a band in both the C and C + T lanes represents a fragment with 3′-C, while a band in only the C + T lane indicates T at the 3′-terminus.

Since the gel can resolve fragments differing by only a single nucleotide and the fragments migrate in inverse order of their sizes, it should be apparent that the shortest fragment, namely *p(Tp) in the C + T lane, will be found closest to the bottom of the gel. The next band up will represent the dinucleotide *pTp(Cp) and will appear in both the C and C + T lanes, while the third lowest band will come from the trinucleotide *pTpCp(Tp) and will appear in only the C + T lane. You should see by now that reading each lane in turn from the bottom of the gel gives you the entire sequence of the DNA sample from the 5′-terminus to the 3′-terminus. The appearance of the gel is diagrammed in Figure 1.

Example 3.1

What is the sequence of the DNA strand represented by each of the Maxam–Gilbert gels shown in Figure 2? (Note that each gel contains the results of two sequencing experiments.)

In each case, the bases are read from the bottom to the top, yielding the sequence in the 5′ to 3′ direction:*

* The base sequences are arranged in triplets for readability only.

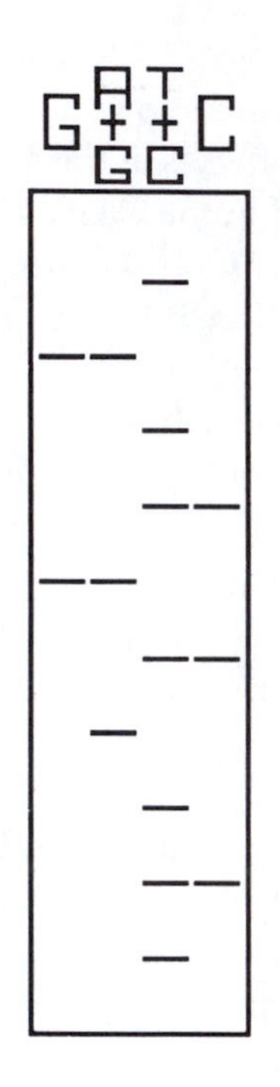

FIGURE 1 Sample Maxam–Gilbert sequencing gel.

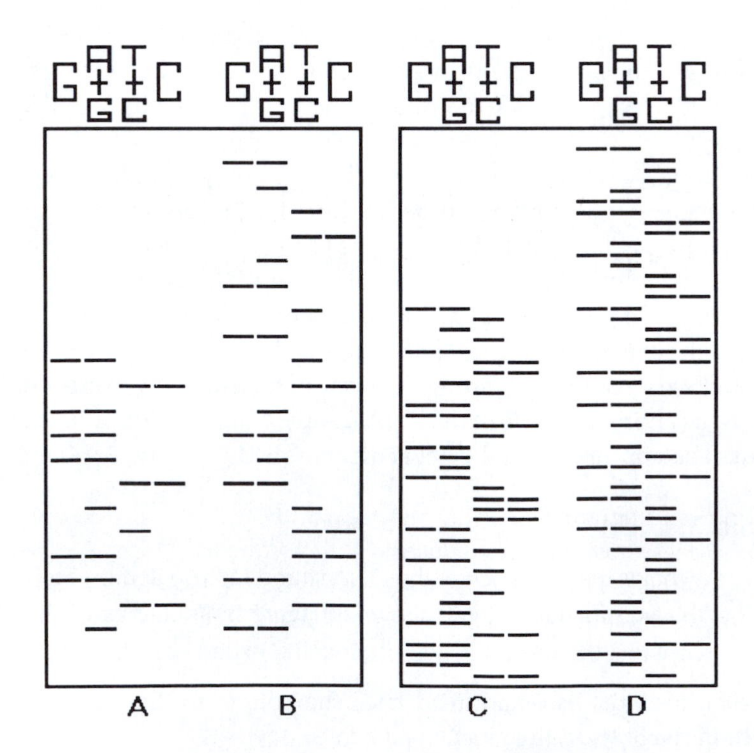

FIGURE 2 Maxam–Gilbert gels for Example 3.1.

A. ACG GTA CGG GCG
B. GCG CCT GAG ACT GTG ACT AG
C. CGG GCG AAT GTG TAG CCC TGA AGT CGG TAC CGT ATG
D. TAG CAT GAT ATT ACG TTG ACG TAT ATG CAG CCC TAG CTT AGA CCG GAT TTG

The Sanger Method

Philosophically, the Maxam–Gilbert method is a degradative one. The Sanger dideoxy procedure, on the other hand, is essentially a synthetic one, in which the abnormally terminated products of DNA polymerase action are evaluated. When polymerase acts on a mixture of all four deoxyribonucleoside-5′-triphosphates in the presence of a DNA template of unknown sequence, it produces the antiparallel complementary DNA strand. If a 2′,3′-dideoxyribonucleoside-5′-triphosphate is present in addition to the normal nucleotide substrates, the enzyme will randomly incorporate it into the growing chain. The result, of course, is termination of that chain, since the possibility of further extension by formation of 3′ → 5′-phosphodiester bonds is eliminated. Under the conditions of the experiment, all possible abnormally terminated chains are produced, yielding a population of DNA molecules of all possible lengths.

Repeating the experiment with a second 2′,3′-dideoxyribonucleoside-5′-triphosphate yields another mixture of chains. In practice four reactions are carried out, each using one of the 2′,3′-dideoxyribonucleoside-5′-triphosphates. Labeling of the 2′,3′-dideoxyribonucleoside-5′-triphosphate in the α position with ^{32}P allows for detection of the fragments on the electrophoresis gel.

If, as we saw in the Maxam–Gilbert case, the template is pTpCpTpApCpGpCpTpGpT, the product mixture in each of the four lanes will be

A lane	C lane	G lane	T lane
ddA*	AddC*	ACAddG*	ACAGCGddT*
ACddA*	ACAGddC*	ACAGCddG*	
ACAGCGTddA*		ACAGCGTAddG*	
ACAGCGTAGddA*			

The autoradiogram will appear as shown in Figure 3. Again, the shortest fragments move most rapidly so that the entire sequence, reading from bottom to top, is ACAGCGTAGA. Note, however, that this is the antiparallel complement of the template whose sequence is TCTACGCTGT.

Example 3.2

What is the sequence of the DNA strand represented by each of the Sanger dideoxy gels shown in Figure 4?

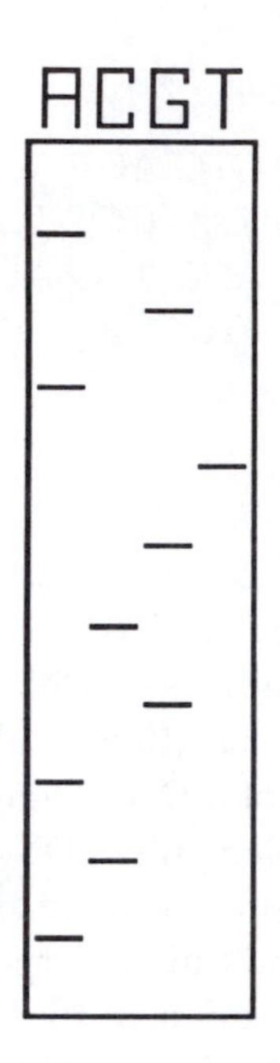

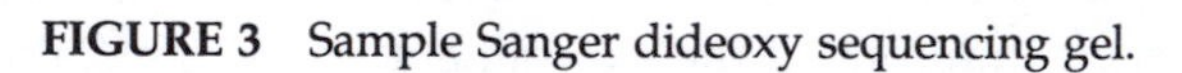

FIGURE 3 Sample Sanger dideoxy sequencing gel.

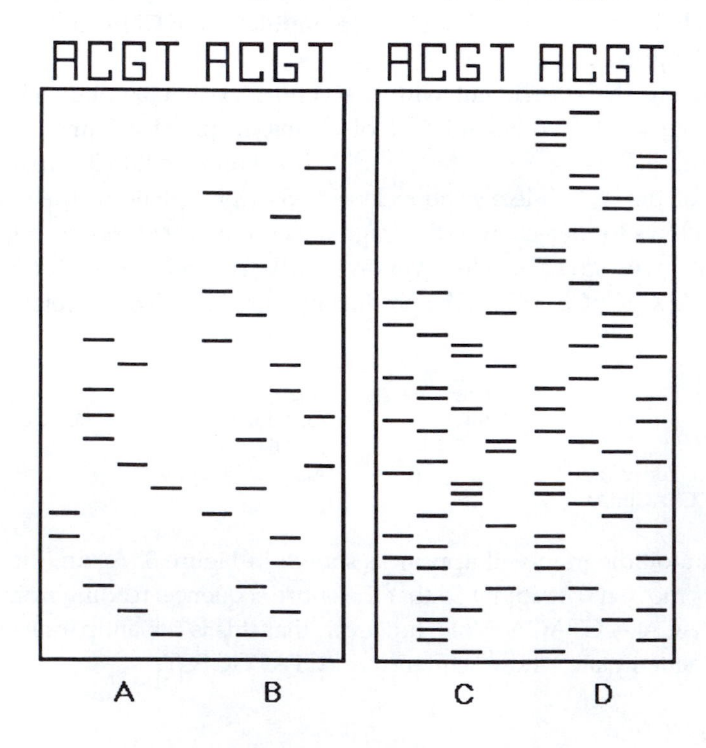

FIGURE 4 Sanger dideoxy gels for Example 3.2.

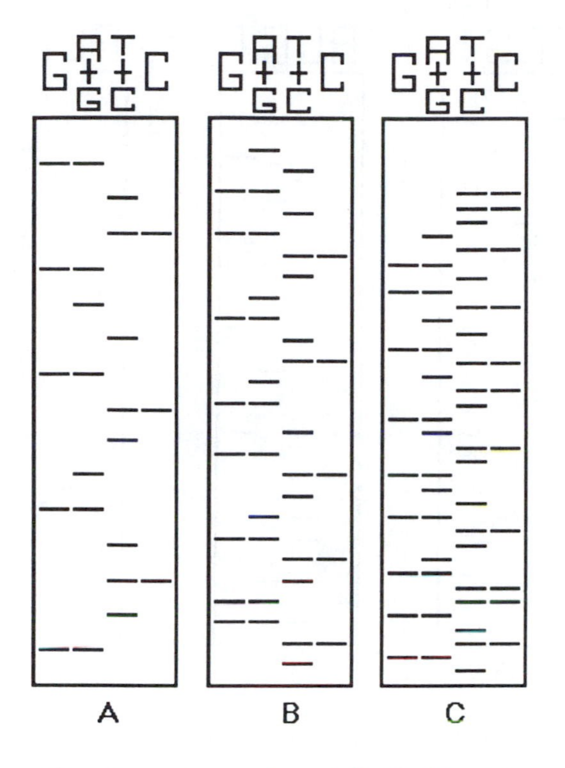

FIGURE 5 Sequencing gels for Problem 3.1.

When the sequences are read from the gels, we get:

A. TGC CAA TGC CCG C
B. CGC GGA CTC TGG ACA CTG ATC
C. GCC CGC TAC ACA TCG GGA CTT CAG CCA TGG CAT AC
D. ATC GTA CTA TAA TGC AAC TGC ATA TAC GTC GGG ATC GAA TCT GGC CTT AAA C

The sequences of the original templates, of course, are the antiparallel complements of those read from the gels, namely:

A. GCG GGC ATT GGC A
B. GAT CAG TGT CCA GAG TCC GCG
C. GTA TGC CAT GGC TGA AGT CCC GAT GTG TAG CGG GC
D. GTT TAA GGC CAG ATT CGA TCC CGA CGT ATA TGC AGT TGC ATT ATA GTA CGA T

End of Chapter Problems

Problem 3.1

What is the sequence of the DNA strand which served as the template for each of the sequencing experiments depicted in Figure 5?

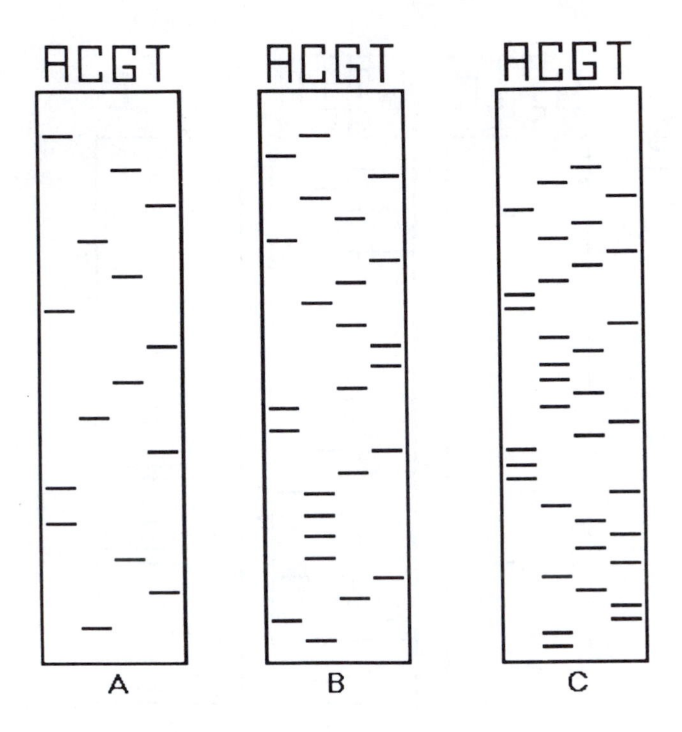

FIGURE 6 Sequencing gels for Problem 3.2.

Problem 3.2

What is the sequence of the DNA strand which served as the template for each of the sequencing experiments depicted in Figure 6?

Analysis of DNA by Restriction Mapping

Introduction

The determination of the base sequence of a large DNA molecule, although it relies on some of the same thought processes as the determination of amino acid sequences, requires a somewhat different approach. Since the "alphabet" of DNA consists of only four letters, one could not hope to develop enzymatic methods that would cleave at specific residues. The result would be hopelessly massive collections of very short oligonucleotides, and reconstructing the original sequence would be impossible. Instead one would seek enzymes that cleaved at specific *base sequences*.

It should be obvious that, given a DNA chain of random base sequence, the probability of finding a given base at any position is 1/4. But the probability of finding a given base *sequence* will be 1/4 raised to the power equal to the number of bases in the sequence. Thus, for example, the probability of finding a particular dinucleotide (of which there are 16) is $(1/4)^2$, or 1/16. The probability of finding a given tetranucleotide is $(1/4)^4$, or 1/256, and the probability of finding a hexanucleotide is $(1/4)^6$, or 1/4096. Since most DNA molecules are double-stranded (we will consider only double-stranded DNA), these same values apply to the probabilities of finding recognition sites of two, four, or six base pairs.

In a double-stranded DNA molecule of 10,000 base pairs (or as the molecular biologists like to say, 10 kilobase pairs, or 10 kbp), a hypothetical enzyme recognizing a single base, for example A, would be expected to cleave once every 4 bases on average, yielding some 2500 products. On the other hand, an enzyme recognizing a tetranucleotide sequence would yield only about 40 fragments (10,000/256), and an enzyme recognizing a 6 base pair sequence would yield only about two or three fragments (10,000/4,096). Such enzymes would make the life of a DNA sequencer much easier.

Nature has apparently recognized the need for enzymes that cleave at specific base sequences. A rudimentary immunologic system has evolved in bacteria, allowing them to destroy the DNA of an invading bacteriophage through the use of restriction endonucleases. These enzymes, of which over 400 have been documented, are found in a wide variety of bacteria. Recognition sites of from four to eight base pairs are known. In virtually every case the recognition site is palindromic, that is, it can be read from either end with the same results. For example, the enzyme EcoR I has the six base pair recognition sequence

5′-GAATTC-3′
3′-CTTAAG-5′

while the recognition sequence for BamH I is

5′-GGATCC-3′
3′-CCTAGG-5′

It is not necessary that each strand of the recognition site be palindromic, only that the base pairs be so. For the EcoR I recognition sequence, whether one reads from left to right or right to left, one encounters base pairs in the order GC, AT, AT, TA, TA, CG.

The name of a restriction endonuclease generally consists of a three letter abbreviation specifying the organism, a strain designation, and a Roman numeral. Fortunately, effective use of these enzymes requires neither an understanding of their names nor memorization of their recognition sequences.

In the process of peptide sequencing it is necessary to cleave the peptide with an enzyme, isolate the fragments (generally using ion exchange chromatography), and determine the amino acid composition of each fragment. By repeating the process with other enzymes one could eventually deduce the entire amino acid sequence and locate the positions of all cleavage sites.

In restriction mapping of DNA, the fragments obtained after enzyme-catalyzed digestion need not be isolated. Instead they are simply separated from each other by electrophoresis on an agarose gel containing the detergent SDS (sodium dodecyl sulfate). The gel acts as a molecular sieve, separating fragments according to molecular size. (This is the same phenomenon that we saw earlier in the discussion of DNA sequencing gels. In that case, however, the fragments to be separated are much smaller and must therefore be resolved on polyacrylamide rather than on agarose.) Fragments are located by staining the gel with the fluorescent dye ethidium bromide and observing the results with an ultraviolet light source. DNA fragments appear as dramatic, intensely colored orange bands against a blue background. The distance migrated by a given band is inversely proportional to the logarithm of the number of base pairs it contains (actually inversely proportional to the logarithm of its molecular weight), so that if a standard mixture containing fragments of known sizes is run on the same gel, in an adjacent lane, it becomes possible to estimate the sizes of the unknown fragments to a reasonable degree of certainty. Repeating the process with one or more enzymes of differing specificity can eventually allow one to deduce the relative position of each of the restriction sites. If further detail is needed, such as the base sequence of the DNA, each of the fragments may be isolated and sequenced according to either the Sanger or the Maxam–Gilbert method.

Single Enzyme Digests

In a sense, restriction mapping may seem less satisfying than peptide sequencing in that "only" the cleavage sites are located. This will, however, prove to be a challenging exercise and will help enable you to use somewhat different logical thought patterns.

Example 4.1

When a 2.0 kbp DNA fragment is digested with EcoR I, fragments of 0.5 and 1.5 kbp are obtained. Show the restriction map.

This is a trivial problem, but it makes an important point. The restriction site is, of course, 0.5 kbp from one end of the molecule — but which end? The answer is that there is no way to know. Since the molecule is double stranded, each end contains a 3′-terminal nucleotide and a 5′-terminal nucleotide. We cannot distinguish one end from the other, so there are, in fact, two equivalent maps which can be drawn, one with the EcoR I site 0.5 kbp from the "left-hand" end of the molecule and one with the cleavage site 0.5 kbp away from the "right-hand" end of the molecule, as shown in Figure 1. (Notice that maps A and B are mirror images of each other.)

Example 4.2

Restriction of a 3.0 kbp DNA sample with Pst I gives fragments of 1.1 and 1.9 kbp. Show the restriction map.

Just as we saw in Example 4.1, there is a single restriction site. Note that the sum of the sizes of the products (1.1 and 1.9 kbp) is equal to the size of the original sample. Again, two equivalent maps (Figure 2) can be shown.

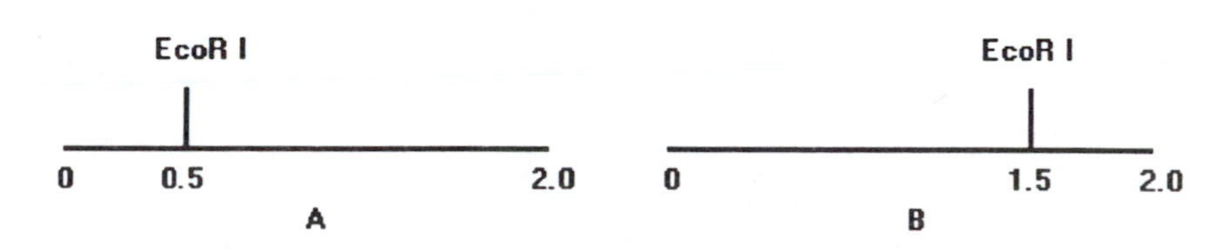

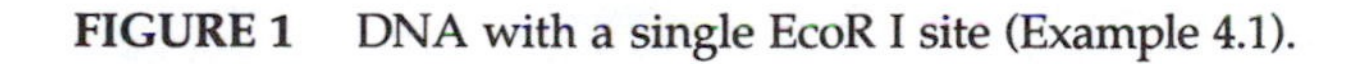

FIGURE 1 DNA with a single EcoR I site (Example 4.1).

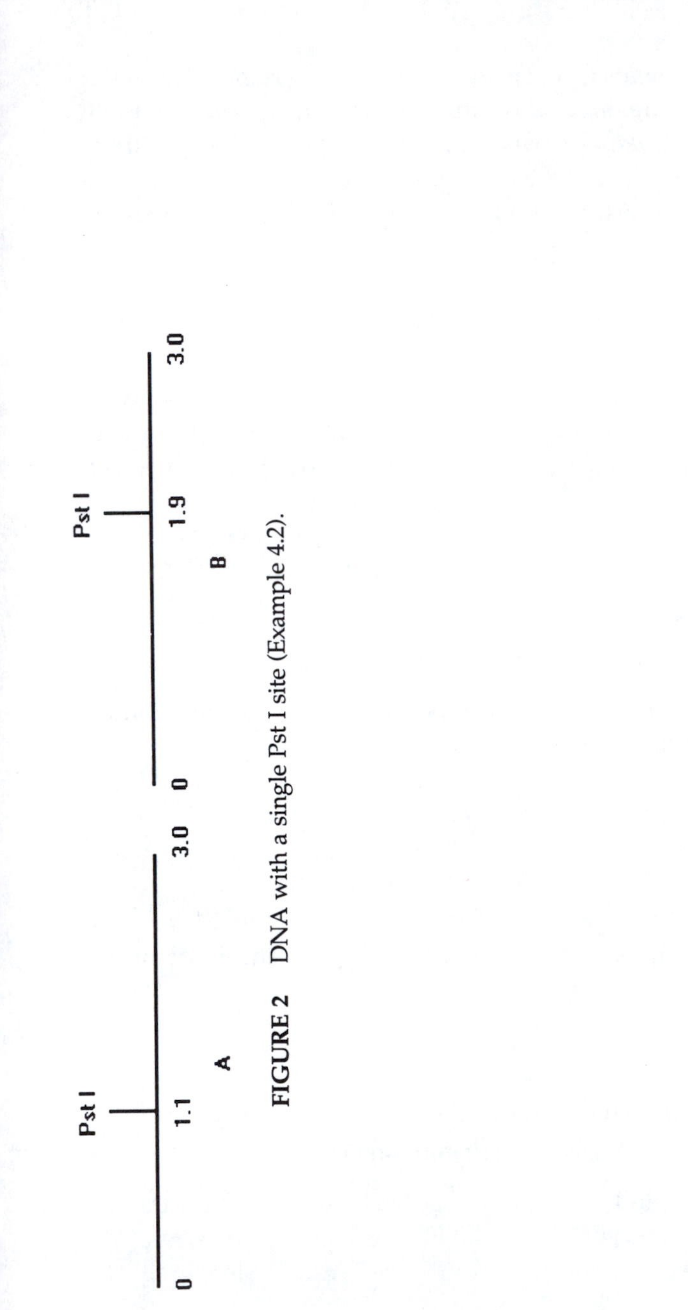

FIGURE 2 DNA with a single Pst I site (Example 4.2).

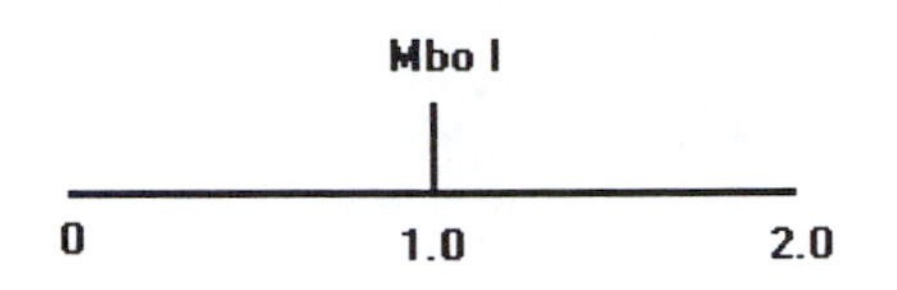

FIGURE 3 Cleavage by Mbo I at midpoint of DNA (Example 4.3).

Example 4.3

Digestion of a 2.0 kbp DNA with Mbo I gives a single band of 1.0 kbp. Show the restriction map.

Cleavage at a single restriction site should yield two fragments (unless, as will be seen later, the original DNA is circular; this complication will not concern us yet). The finding of a single band whose size is exactly half that of the original DNA suggests, of course, that the restriction site is right in the middle of the molecule, at 1.0 kbp from either end (Figure 3).

Example 4.4

When a 3.0 kbp DNA fragment is digested with BamH I, three products are obtained. Their sizes are 0.5, 1.0, and 1.5 kbp. Show the restriction map.

When more than two restriction fragments are obtained the number of possible restriction maps increases. The best procedure is to draw *all* possible maps, then eliminate those that are duplicates of others. In this problem, we should recognize that there are two restriction sites yielding three products, and we should show a total of six possible maps.

The way in which the six maps are derived is instructive. Start with the smallest fragment (0.5 kbp). If that fragment is closest to the "left" side of the molecule, then a single cleavage further toward the "right" will yield the remaining two fragments. The second cleavage can be either 1.0 or 1.5 kbp away from the first. Thus two possible maps have been generated.

Repeat the procedure, but this time place the second smallest fragment at the "left" end, generating two more possible maps. Since the first cleavage site is now shown at 1.0 kbp, the second will be at either 1.5 or 2.5 kbp.

The two final maps can be obtained by placing the largest fragment at the "left" (cleavage at 1.5 kbp) and locating the second cleavage site at either 2.0 or 2.5 kbp. The results are shown in Figure 4.

In order to eliminate duplicate maps, list each map by fragment length from "left" to "right":

A: 0.5, 1.0, 1.5
B: 0.5, 1.5, 1.0
C: 1.0, 0.5, 1.5

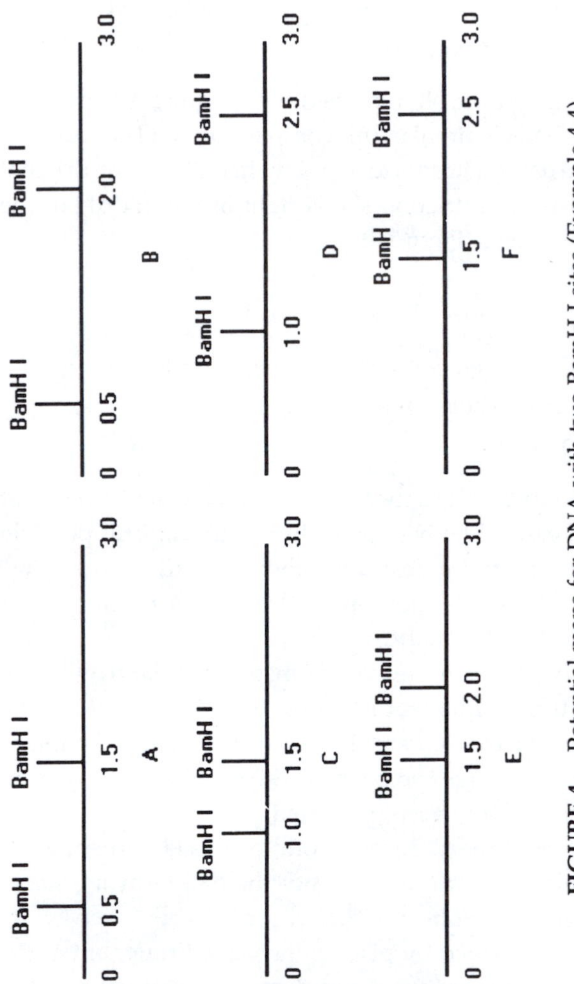

FIGURE 4 Potential maps for DNA with two BamH I sites (Example 4.4).

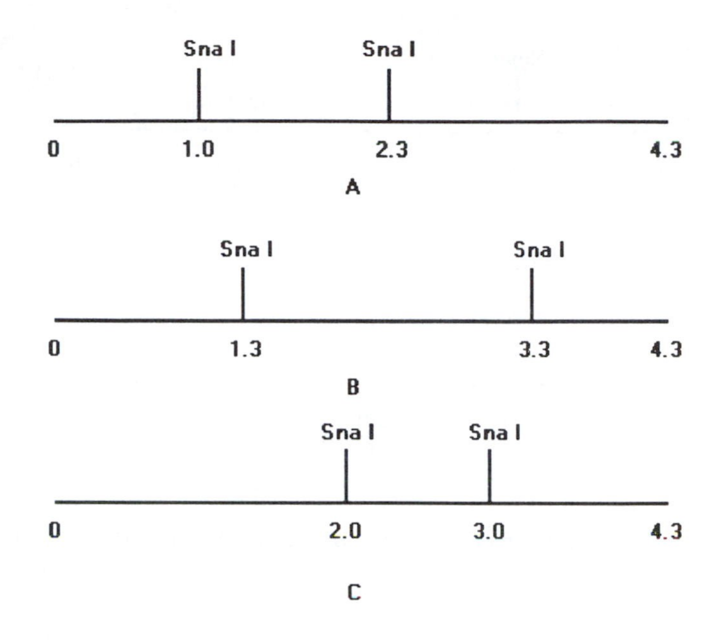

FIGURE 5 Potential maps for DNA with two Sna I sites (Example 4.5).

D: 1.0, 1.5, 0.5
E: 1.5, 0.5, 1.0
F: 1.5, 1.0, 0.5

We can then group the maps in pairs, such that each member of the pair differs from the other by rotation through 180°. The pairs are

A: 0.5, 1.0, 1.5	F: 1.5, 1.0, 0.5
B: 0.5, 1.5, 1.0	D: 1.0, 1.5, 0.5
C: 1.0, 0.5, 1.5	E: 1.5, 0.5, 1.0

A and F are identical, as are B and D, and C and E. Thus there are only three unique possibilities for a linear DNA with two cleavage sites. Additional data are needed to solve this problem. In general, it will not be necessary to draw all possible restriction maps, since we will develop analytical methods to avoid this trial and error approach. There may be times, however, when you will want to use this approach to check your conclusions.

Example 4.5

Restriction of a 4.3 kbp DNA with Sna I gives fragments of 1.0, 1.3, and 2.0 kbp. Show the restriction maps.

As in the preceding example, production of three fragments implies two cleavage sites. There are three possible restriction maps, as shown in Figure 5.

The maps you have shown may be the mirror images of any or all of these. As we have seen, in the case of multiple cleavage sites for a single restriction enzyme, mirror image maps are identical.

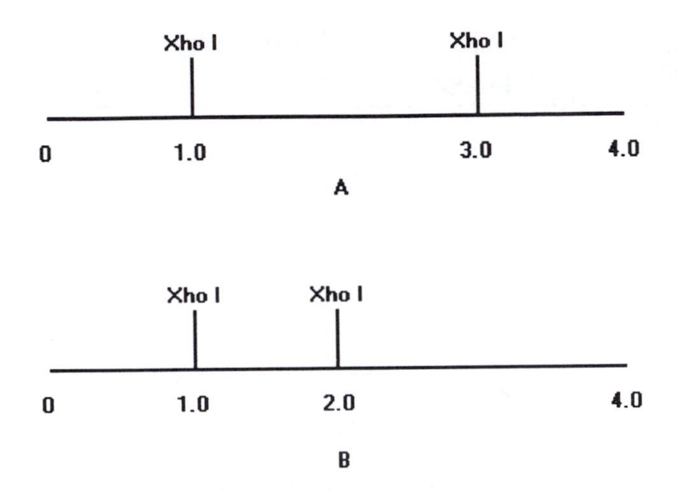

FIGURE 6. Potential maps for DNA with two Xho I sites (Example 4.6).

Example 4.6

Electrophoresis of the Xho I restriction fragments of a 4.0 kbp DNA gives bands of 1.0 and 2.0 kbp. Show the possible restriction maps.

As in Example 4.3, the sum of the sizes of the restriction fragments is less than the size of the original DNA. This suggests the presence of multiple bands of a given size, in this case, two bands of 1.0 kbp and a single band of 2.0 kbp. Only two restriction maps are possible (Figure 6).

Example 4.7

How many distinct restriction maps can be shown for a DNA with three restriction sites for a given enzyme?

The number of possible restriction maps increases rapidly as the number of restriction sites increases. When there is one restriction site, two possible maps can be drawn. These have been shown to be identical. For a substrate with two restriction sites, a total of six maps can be shown, which reduces to three unique possibilities. For a DNA with three restriction sites, as in this example, we can initially show a total of 24 possible maps. Allowing for duplication, we will end up with 12 unique maps.

Mathematically, the total number of possible maps is N!, where N is the number of restriction fragments generated. Since the collection of maps always resolves itself into duplicate pairs, the number of unique maps is N!/2. This value must decrease, of course, when there are multiple fragments of the same size (see Example 4.6). It should be apparent that even a relatively low number of restriction sites leads to collections of potential maps that can be overwhelming. The problem is to find a means of eliminating most of the possibilities and, ultimately, finding the one map that correctly describes the molecule under consideration. We will see how this is done in the next few examples as we explore the concept of multiple digests in which two (or more) restriction enzymes are utilized.

Multiple Enzyme Digests

Just as complete determination of the sequence of amino acids in a protein usually requires several cleavage methods, determination of a restriction map of DNA generally requires at least two restriction enzymes. Whether the enzymes are used simultaneously or sequentially makes no difference for us, since the products will be the same. In actual laboratory work, however, most multiple digests will be carried out sequentially, since the buffer requirements of the enzymes usually differ.

Example 4.8

When a 3.0 kbp DNA fragment is digested with BamH I, fragments of 0.5, 1.0, and 1.5 kbp are obtained. When the same DNA is digested with EcoR I, fragments of 1.2 and 1.8 kbp are obtained. Redigestion of the BamH I digest with EcoR I gives fragments of 0.5, 0.7, 0.8, and 1.0 kbp. Show the restriction map.

There is one restriction site for EcoR I; we can place it either at 1.2 or 1.8 kbp. Let us arbitrarily choose the site at 1.2 kbp.

BamH I produces three fragments, indicating two restriction sites. When we look at the double digest data, we see that the EcoR I fragment of 1.2 kbp is replaced by two fragments of 0.5 and 0.7 kbp, respectively. Since this EcoR I fragment is the result of cleavage at a position 1.2 kbp from the "left" end of the molecule, the BamH I site within this fragment must be either at 0.5 kbp or at 0.7 kbp.

The 1.8 kbp EcoR I fragment is replaced (in the double digest) by fragments of 0.8 and 1.0 kbp. Thus the second BamH I site must be either 0.8 or 1.0 kbp away from the EcoR I site, i.e., either at 2.0 kbp or at 2.2 kbp.

With the EcoR I site at 1.2 kbp, the BamH I sites are at either 0.5 or 0.7 kbp, and at either 2.0 or 2.2 kbp.

When BamH I acts alone, three fragments are produced. Two of them, namely the 0.5 kbp and the 1.0 kbp fragments, are also found in the double digest. Thus we can say that EcoR I cleaves only the 1.5 kbp BamH I fragment, yielding fragments of 0.7 and 0.8 kbp. Consequently, BamH I must cleave on both sides of the EcoR I site, at positions 0.5 and 2.0 kbp or at positions 0.4 and 1.9 kbp. Of the two possibilities, only cleavage at 0.5 and 2.0 kbp is consistent with our previous analysis. The resulting restriction map is shown in Figure 7.

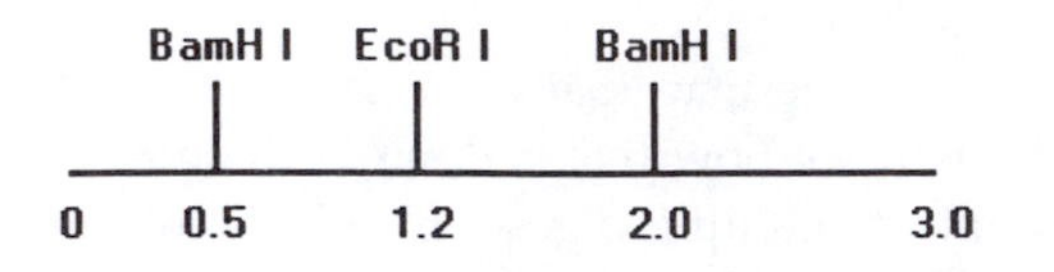

FIGURE 7 EcoR I/BamH I double digest (Example 4.8).

Example 4.9

When a 3.0 kbp DNA fragment is digested with BamH I, fragments of 0.5, 1.0, and 1.5 kbp are obtained. When the same DNA is digested with EcoR I, fragments of 0.7 and 2.3 kbp are obtained. When the BamH I digest is redigested with EcoR I, fragments of 0.5, 0.7, 0.8, and 1.0 kbp are obtained. Show the restriction map.

The data for this example look remarkably similar to those in Example 4.8. There is, however, an important difference.

We can take the EcoR I site as 2.3 kbp. The double digest data indicate that the small EcoR I fragment is not further digested by BamH I, since a 0.7 kbp fragment survives in the double digest. The 2.3 kbp EcoR I fragment is cleaved twice by BamH I, yielding fragments of 0.5, 0.8, and 1.0 kbp.

Similarly we can say that of the three BamH I fragments, only the 1.5 kbp fragment is further cleaved by EcoR I. That is, the EcoR I site is located within the 1.5 kbp BamH I fragment. You should be able to see, then, that the 1.5 kbp BamH I fragment must include an end of the DNA molecule, i.e., must cover the range 1.5 to 3.0 kbp.

The remaining BamH I site is either 0.5 or 1.0 kbp away from the first one, i.e., at either 1.0 or 0.5 kbp. The data do not allow a decision, so that we must show two possible restriction maps (Figure 8).

The difference between this problem and the previous one is, of course, the relative locations of the restriction sites. In general we will see that a second restriction enzyme cleaving between two sites of the first enzyme will provide more information than one cleaving "outside" those two sites.

In some applications the unknown DNA is labeled at both 5′-ends by treating with polynucleotide kinase in the presence of [^{32}P]-ATP. This end-labeled fragment is then restricted with one or more endonucleases, either

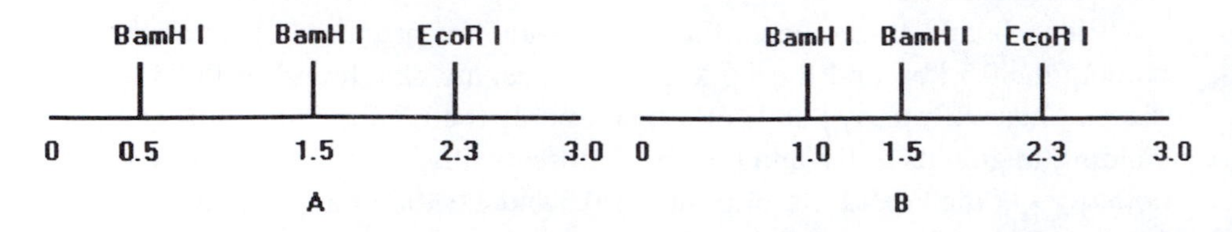

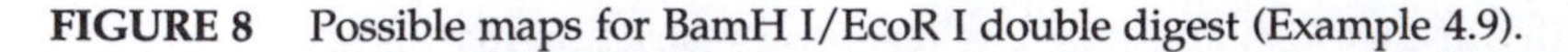

FIGURE 8 Possible maps for BamH I/EcoR I double digest (Example 4.9).

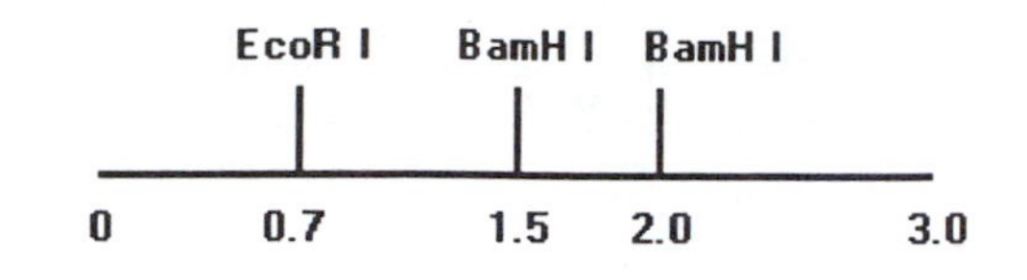

FIGURE 9 BamH I/EcoR I double digest (Example 4.10).

singly or in pairs. Fragments are separated by electrophoresis on agarose gels and visualized by autoradiography. It is important to remember that in this procedure only the ends of the DNA are seen. Consequently, the sum of the lengths of the fragments will often not be identical to the length of the original DNA. An experiment of this type may complement a nonradioactive experiment, or completely supplant it.

Example 4.10

A 3.0 kbp DNA labeled with ^{32}P at both 5′-ends, is restricted with BamH I and EcoR I, both singly and together. The autoradiogram showed the following bands:

BamH I	1.0, 1.5 kbp
EcoR I	0.7, 2.3 kbp
BamH I/EcoR I	0.7, 1.0 kbp.

Show the restriction map.

The two EcoR I fragment sizes add up to 3.0 kbp, indicating that there is only a single EcoR I site which we can arbitrarily place at 0.7 kbp. The two BamH I fragments, however, total only 2.5 kbp, suggesting two BamH I sites separated by 0.5 kbp. Remember that the central 0.5 kbp fragment, which is not labeled, will not be seen. Since the double digest contains a 1.0 kbp fragment, the EcoR I site at 0.7 kbp must be within the 1.5 kbp BamH I fragment. The first BamH I site, therefore, must be at position 1.5 kbp, and the second site, 0.5 kbp away, must be at 2.0 kbp, as shown in Figure 9.

You should have recognized that this problem is virtually identical to that presented in Example 4.9. In that problem we wound up with two equally probable maps; knowing which fragments arose from the ends of the DNA molecule would have eliminated one of the possibilities. In Example 4.10 the analysis is much easier and more complete.

Example 4.11

When restriction mapping is carried out in the laboratory, fragment mixtures are resolved by electrophoresis on agarose gels. Typically a standard fragment mixture is run in one lane of the gel, and the results of various digests are run in others. Since the

FIGURE 10 Dra I/Bst I double digest electrophoresis gel (Example 4.11).

Table 1

Restriction Fragment Standards for Electrophoresis[a]

pGEM DNA Markers		1 Kb Ladder	
36	396	75	2036
51	460	134	3054
65	517	154	4072
75	676	201	5090
126	1198	220	6018
179	1605	298	7126
222	2645	344	8144
350		396	9162
		506	10180
		517	11198
		1018	12216
		1636	

[a] All fragment sizes in base pairs.

sizes (in kbp) of the standard fragments are known, the sizes of the unknown fragments can be estimated by comparison.

The gel depicted in Figure 10 shows the restriction fragments obtained in a double digest experiment. Wells 1 and 2 contain a Dra I digest and a Bst I digest, respectively. Well 3 contains a Dra I/Bst I double digest, and well 4 contains the pGEM® (Promega Corp.) DNA Markers, whose sizes are listed in Table 1.

Table 2

Electrophoretic Migratory Distances of pGEM DNA Markers

Base pairs	Log (bp)	Distance (mm)
36	1.556	64.7
51	1.708	59.7
65	1.813	56.0
75	1.875	53.9
126	2.100	46.8
179	2.253	41.3
222	2.346	38.6
350	2.544	32.2
396	2.598	29.8
460	2.663	27.6
517	2.713	26.1
676	2.830	22.1
1198	3.078	14.0
1605	3.205	9.9
2645	3.422	2.8

By measuring the distance from the bottom of well 4 to each of the bands, you should be able to draw a graph that relates migratory distance to fragment size for the standards. That, in turn, will allow you to calculate the sizes of the bands in the first three wells. It is important to take the time to measure the distances carefully. Use a good ruler and measure to the nearest 0.1 mm. After a while this will become quite tedious, but the results will be well worth the effort you will expend. It would be wise not to proceed in this text until you have measured the gel and plotted your data.

The fragment sizes, in base pairs, for the standard mixture, their logarithms, and the measured distances of the electrophoretic bands are listed in Table 2.

When electrophoretic migratory distance is plotted as a function of fragment base pair length, the graph shown in Figure 11 is obtained. This is, of course, an exponential curve and, being nonlinear, is not very useful.

It is only when the distance is plotted as a function of the logarithm of the number of base pairs, as in Figure 12, that we get a straight line whose slope and intercept can be readily calculated:

$$\text{slope} = -33.2$$
$$\text{intercept} = 116$$

The equation for this line is

$$\text{distance} = -33.2\ (\log \text{bp}) + 116$$

or

$$(\log \text{bp}) = (116 - \text{distance})/33.2$$

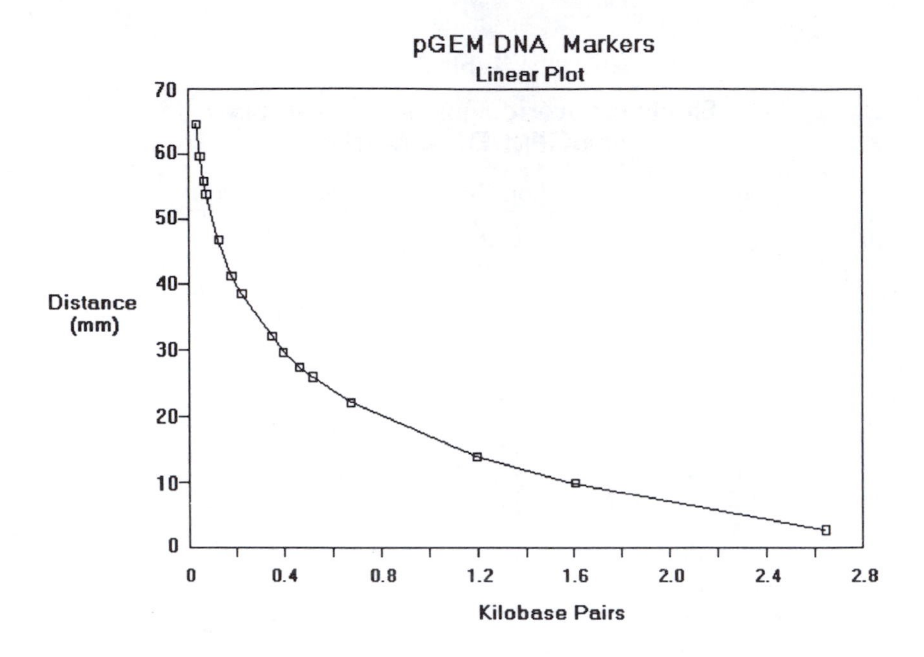

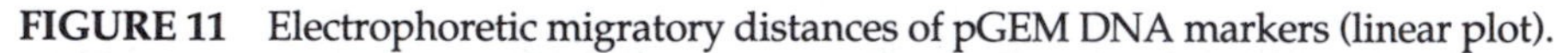

FIGURE 11 Electrophoretic migratory distances of pGEM DNA markers (linear plot).

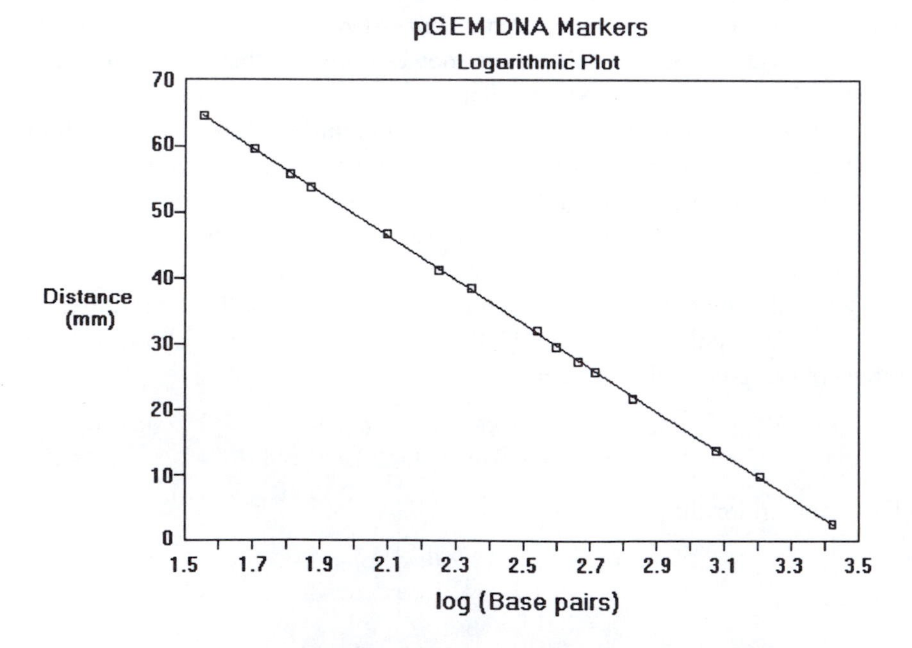

FIGURE 12 Electrophoretic migratory distances of pGEM DNA markers (logarithmic plot).

Table 3

Dra I/Bst I Double Digest: Calculated Sizes of DNA Fragments

Lane 1 Dra I		Lane 2 Bst I		Lane 3 Dra I/Bst I	
Distance*	kb	Distance	kb	Distance	kb
15.1	1.1	13.0	1.3	20.0	0.8
20.0	0.8	14.2	1.2	21.4	0.7
24.0	0.6			23.8	0.6
				29.6	0.4
Sum	2.5		2.5		2.5

* All distances are in millimeters.

or

$$\text{base pairs} = \text{antilog}\ [(116 - \text{distance})/33.2]$$

Thus, to calculate the size of any fragment, subtract its measured migratory distance (in mm) from 116, divide the difference by 33.2, and take the antilog of the result. In this problem, the calculated values of the fragment sizes are shown in Table 3.

Now, applying the kind of analysis we have used before, we can assign the Bst I cleavage site to position 1.2 kbp. In the double digest the 1.1 kbp Dra I fragment is cleaved by Bst I to give fragments of 0.4 and 0.7 kbp, so we conclude that the Dra I sites are 0.4 and 0.7 kbp away from the Bst I site, i.e., at positions 0.8 and 1.9 or at positions 0.5 and 1.6. Cleavage at 0.5 and 1.6 kbp would yield fragments of 0.5 and 1.9 kbp, neither of which is found. Thus, only cleavage at 0.8 and 1.9 kbp is consistent with the data from the Dra I digest. The entire map is as shown in Figure 13.

End of Chapter Problems

Problem 4.1

Ava I acting on a 2.8 kbp DNA gives fragments of 1.0 and 1.8 kbp. Bgl II acting on the same DNA gives fragments of 0.3 and 2.5 kbp. A double digest using

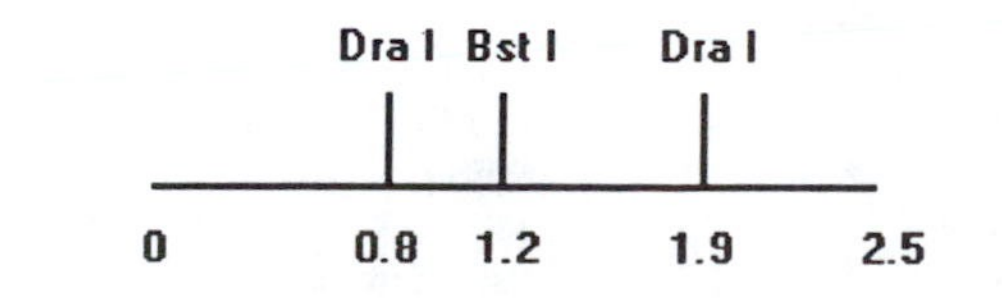

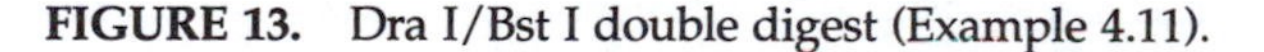

FIGURE 13. Dra I/Bst I double digest (Example 4.11).

both enzymes gives fragments of 0.3, 0.7, and 1.8 kbp. Draw the restriction map.

Problem 4.2

A Sno I digest of a 2.6 kbp DNA gives fragments of 0.7, 0.8, and 1.1 kbp. A Cfo I digest of the same DNA gives fragments of 1.0 and 1.6 kbp. A Sno I/Cfo I double digest gives fragments of 0.3, 0.5, 0.7, and 1.1 kbp. Show the restriction map of the DNA.

Problem 4.3

Restriction of a 3.5 kbp DNA with Pst I gives fragments of 0.5, 0.7, 0.9, and 1.4 kbp. Restriction of the same DNA with Hind III gives fragments of 0.8 and 2.7 kbp. A Pst I/Hind III double digest gives fragments of 0.3, 0.4, 0.5, 0.9, and 1.4 kbp. Show the restriction map.

Problem 4.4

A 3.0 kbp DNA is digested with three different restriction endonucleases, both singly and in pairs. The data are listed in Table 4. Show the restriction map.

Table 4

Data For Problem 4-4

Enzyme	Fragments (kbp)			
Hinf I	0.5	1.0	1.5	
AtuC I	0.7	2.3		
Gsb I	0.3	2.7		
Hinf I/AtuC I	0.5	0.7	0.8	1.0
Hinf I/Gsb I	0.2	0.3	1.0	1.5

Problem 4.5

A 5.0 kbp DNA labeled at both 5′-ends with ^{32}P is restricted with five different enzymes, both singly and in selected pairs. The results are shown in Table 5. Draw the restriction map of the 5.0 kbp DNA and predict the results of the following multiple digests on this same substrate:

Table 5

Data For Problem 4.5

Enzyme	Fragment 1[a]	Fragment 2	Enzyme	Fragment 1	Fragment 2
BamH I	1.2	3.8	SmaI/EcoR I	0.9	2.3
EcoR I	0.9	4.1	EcoR I/BamH I	0.9	3.8
Mbo I	2.2	2.8	Mbo I/BamH I	1.2	2.8
Sma I	2.3	2.7	Mbo I/Xho I	0.8	2.2
Pst I	1.9	3.1	Pst I/Sma I	1.9	2.7
Xho I	0.8	4.2			

[a] All fragment lengths in kbp.

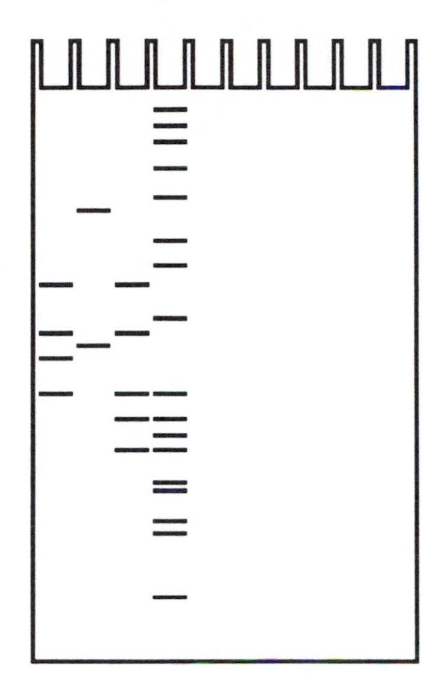

FIGURE 14 Pst I/Hind III double digest electrophoresis gel (Problem 4.7).

A. EcoR I/Xho I
B. BamH I/Pst I
C. BamH I/Sma I/Pst I

Problem 4.6

A 5.30 kbp circular double-stranded DNA is digested with three different restriction endonucleases. The products of digestion (in kbp) are:

EcoR I	5.30
Hpa I	5.30
Alu I	0.75, 2.00, 2.55

The Alu I fragments are isolated and individually digested with EcoR I. The only one cleaved is the 2.00 kbp fragment, which gives fragments of 0.50 and 1.50 kbp.

Digestion of the EcoR I product with Hpa I gives two fragments of lengths 1.30 and 4.00 kbp. When the 1.30 kbp fragment is further digested with Alu I, fragments of 0.50 and 0.80 kbp are obtained. Draw the complete restriction map.

Problem 4.7

The data in Problem 4.3 were calculated from the gel shown in Figure 14. Well 4 contains the GIBCO BRL 1 Kb Ladder DNA standards (fragment sizes are listed in Table 1). Only fragments smaller than 8 kbp are shown. Note, however, that only 17 bands appear. The two bands of 506 and 517 kbp

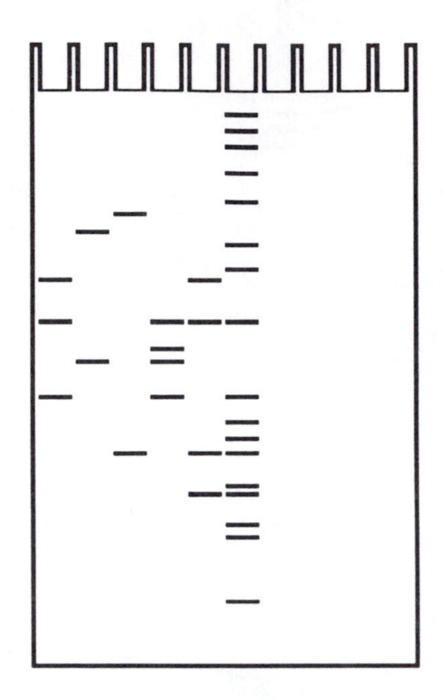

FIGURE 15 Hinf I/AtuC I/Gsb I multiple digest electrophoresis gel (Problem 4.8).

appear at the same position and are seen as a single band. Plot the data and see if you get the fragment pattern shown in Problem 4.3.

Problem 4.8

The gel shown in Figure 15 is the basis of Problem 4.4. Measure the migration distances to determine the lengths of the various fragments. Wells 1 to 3 contain DNA digested with Hinf I, AtuC I, and Gsb I, respectively. Well 4 contains a Hinf I/AtuC I double digest, and well 5 contains a Hinf I/Gsb I double digest. The standard (well 6), again, is the 1 Kb DNA Ladder.

Problem 4.9

Digestion of DNA with Nar I and Sac I, both individually and in combination, gives the agarose gel pattern shown in Figure 16. Nar I and Sac I digests are in wells 1 and 2, respectively, and a Nar I/Sac I double digest is in well 3. The 1 kb DNA Ladder is in well 4. Deduce the restriction map of the DNA.

Problem 4.10

When a sample of DNA is digested with Mla I and Spc I, the gel shown in Figure 17 is obtained. Wells 1, 2, and 3 contain digests using Mla I, Spc I, and a combination of Mla I and Spc I, respectively. The standard in this experiment (well 4) is the pGEM DNA Markers. Deduce the restriction map.

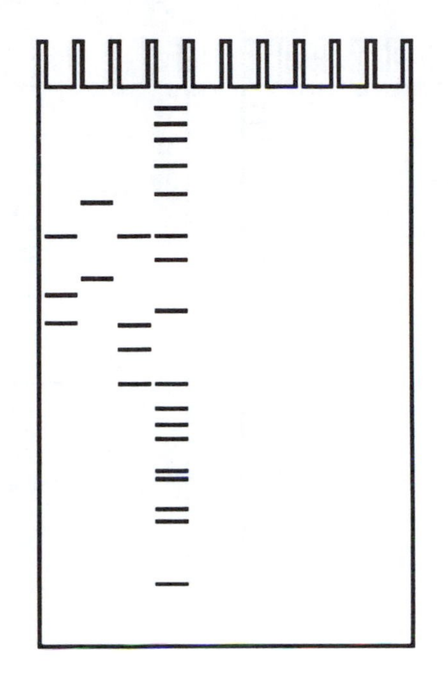

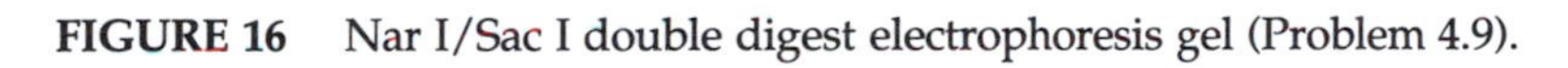

FIGURE 16 Nar I/Sac I double digest electrophoresis gel (Problem 4.9).

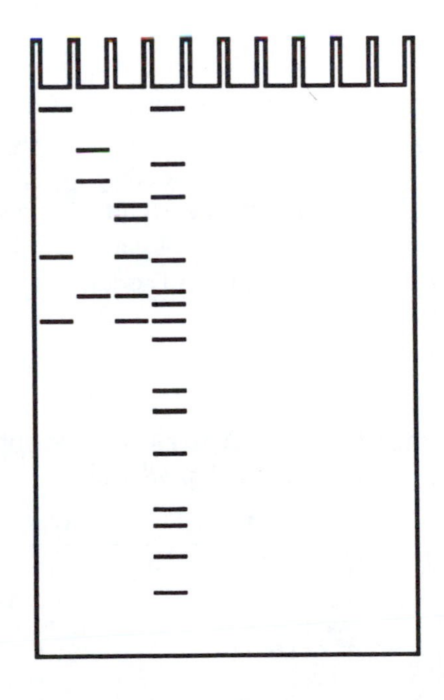

FIGURE 17 Mla I/Spc I double digest electrophoresis gel (Problem 4.10).

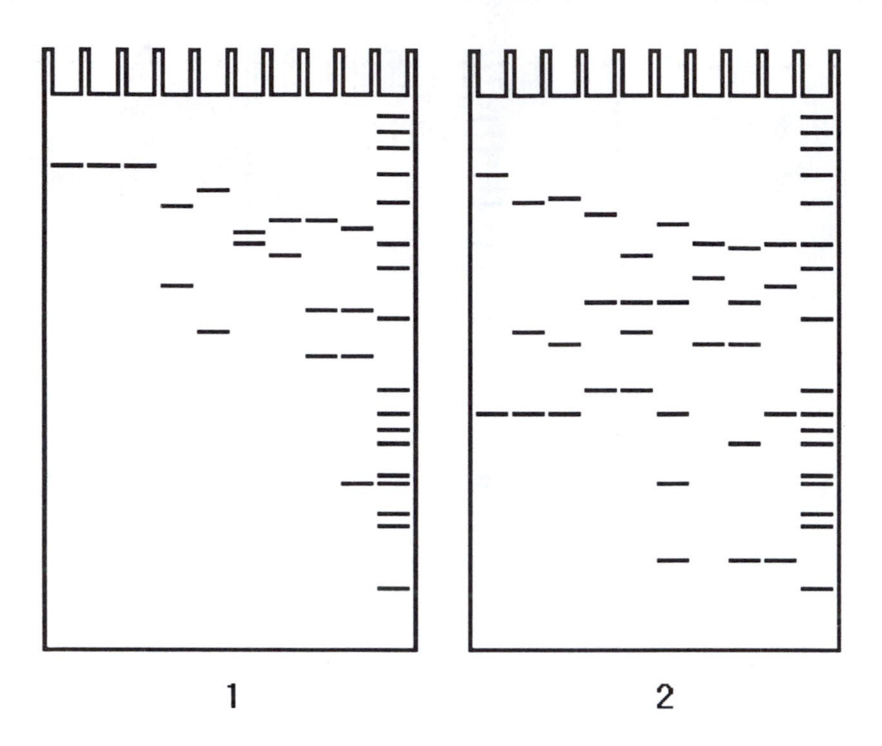

FIGURE 18 Plasmid pBR322 restriction mapping gels (Problem 4.11).

Problem 4.11

A restriction mapping experiment is carried out on a sample of plasmid pBR322 DNA. Eighteen different digests are employed, using from one to three enzymes each. When the restriction digests are analyzed electrophoretically, the gels shown in Figure 18 are obtained. The enzymes used for the samples in each of the various lanes are listed in Table 6. The standard for both gels is the 1 Kb DNA Ladder. Show the restriction map of plasmid pBR322.

Table 6

Enzymes Used in Restriction Mapping of Plasmid pBR322

	Gel 1	Gel 2
Lane 1	EcoR I	Drd I
Lane 2	Bal I	Drd I/Ase I
Lane 3	Ase I	Drd I/Bal I/
Lane 4	EcoR I/Bal I	Sno I
Lane 5	EcoR I/Ase I	Sno I/Bal I
Lane 6	Bal I/Ase I	Sno I/Ear I
Lane 7	Ear I	Rsa I
Lane 8	Ear I/Ase I	Rsa I/Drd I
Lane 9	EcoR I/Ase I/Ear I	Rsa I/Ear I

Additional Problems

Problem 5.1

Octapeptide A has the composition (Ala,Asp,Glu,Gly,Leu,Lys_2,Phe). Treatment of A with $LiBH_4$ followed by hydrolysis gives 2-amino-1,5-pentanediol; treatment with FDNB followed by hydrolysis gives DNP–lysine.

Tryptic digestion of the octapeptide gives three fragments, whose compositions are:

T1 = Lys
T2 = (Lys,Leu)
T3 = (Ala,Asp,Glu,Gly,Phe)

Treatment of T3 with FDNB, followed by hydrolysis gives DNP–aspartate. Chymotryptic digestion of A gives two products, with the compositions:

CT1 = (Ala,Glu)
CT2 = (Asp,Gly,Leu,Lys_2,Phe)

Give the amino acid sequence of A.

Problem 5.2

Decapeptide B had the composition (Arg,Lys,Met,Pro,Ser,Thr,Tyr,Val_3). Treatment of B with FDNB followed by hydrolysis gives DNP–proline. Treatment with lithium borohydride followed by hydrolysis gives 2-amino-1,3-propanediol. Tryptic digestion of B gives three products whose compositions are:

T1 = (Ser,Thr)
T2 = (Lys,Pro,Val)
T3 = (Arg,Met,Tyr,Val_2)

Chymotryptic digestion of B gives two fragments; their amino acid compositions are:

CT1 = (Lys,Met,Pro,Tyr,Val)
CT2 = (Arg,Ser,Thr,Val_2)

What is the amino acid sequence of B?

Problem 5.3

The amino acid composition of decapeptide C is

(Arg,Leu,Lys,Met,Phe_2, Ser,Thr,Tyr,Val)

End group analysis gives DNP–lysine upon treatment with FDNB and hydrolysis, and 2-amino-1,3-butanediol after lithium borohydride reduction and hydrolysis.

Digestion of C with trypsin gives three fragments with the compositions:

T1 = Lys
T2 = (Leu,Phe,Thr)
T3 = (Arg,Met,Phe,Ser,Tyr,Val)

Digestion of C with chymotrypsin gives four fragments with the compositions:

CT1 = Thr
CT2 = (Phe,Lys)
CT3 = (Met,Ser,Tyr)
T4 = (Arg,Leu,Phe,Val)

Treatment of C with cyanogen bromide gives two products, with the compositions:

CB1 = (Hsr,Lys,Phe,Ser)
CB2 = (Arg,Leu,Phe,Thr,Tyr,Val)

What is the amino acid sequence of C?

Problem 5.4

Peptide D has the amino acid composition

(Ala,Arg_2,Glu_3,Lys_2,Met,Phe,Pro,Ser_2, Tyr,Val)

Treatment of D with CNBr gives two products with the compositions:

CB1 = (Ala,Arg,Glu,Lys_2,Phe,Pro)
CB2 = (Arg,Glu_2,Hsr,Ser_2,Tyr,Val)

Digestion of D with chymotrypsin gives three fragments whose amino acid compositions are:

CT1 = (Arg,Glu,Ser,Tyr)
CT2 = (Glu,Met,Phe,Ser,Val)
CT3 = (Ala,Arg,Glu,Lys_2,Pro)

Tryptic digestion of D gives four products with the compositions:

T1 = Arg
T2 = (Glu,Lys,Pro)
T3 = (Arg,Glu,Ser)
T4 = (Ala,Glu,Lys,Met,Phe,Ser,Tyr,Val)

When D is digested with pepsin, five fragments are obtained. Their amino acid compositions are:

P1 = (Ser,Tyr)
P2 = (Glu,Lys,Pro)
P3 = (Arg,Glu,Ser)
P4 = (Glu,Met,Val)
P5 = (Ala,Arg,Lys,Phe)

Give the complete amino acid sequence of peptide D, together with enough of your reasoning to justify your answer.

Problem 5.5

Peptide E has the amino acid composition

$$(Ala_3,Arg_2,Asp,Glu,Gly,His,Ile,Leu,Phe_2,Ser,Tyr,Val_2)$$

Treatment of E with trypsin gives three fragments whose compositions are:

T1 = (Ala,His,Phe)
T2 = (Arg,Gly,Leu,Phe,Val)
T3 = $(Ala_2,Arg,Asp,Glu,Ile,Ser,Tyr,Val)$

When E is treated with chymotrypsin three products are obtained. Their compositions are:

CT1 = (Asp,Arg,Gly,Phe,Val)
CT2 = (Ala,Arg,His,Leu,Phe,Val)
CT3 = (Ala_2,Glu,Ile,Ser,Tyr)

A peptic digest of E contains six products:

P1 = Phe
P2 = (Ala,Ile)
P3 = (Tyr,Val)
P4 = (Ala,Glu,Ser)
P5 = (Arg,Asp,Gly)
P6 = (Ala,Arg,His,Leu,Val)

Treatment of fragment P2, or P4, or P6 with lithium borohydride followed by hydrolysis gives 2-amino-1-propanol. Give the amino acid sequence of E together with enough of your reasoning to justify your answer.

Problem 5.6

Peptide F has the amino acid composition

$$(Arg_3,Asp,Cys,Glu,Gly,Ile_2,Leu_2,Phe,Pro,Ser_2,Tyr,Val)$$

Treatment of F with lithium borohydride followed by hydrolysis gives 3-amino-4-hydroxybutanoic acid.

Tryptic digestion of F gives four products, with the compositions:

T1 = (Arg,Ile)
T2 = (Arg,Val)
T3 = (Arg,Glu,Gly,Ser,Tyr)
T4 = (Asp,Cys,Ile,Leu_2,Phe,Pro,Ser)

Digestion of F with chymotrypsin gives three fragments, whose amino acid compositions are:

CT1 = (Asp,Cys,Leu,Ser)
CT2 = (Arg,Glu,Ile,Ser,Tyr)
CT3 = (Arg_2,Gly,Ile,Leu,Phe,Pro,Val)

When F is digested with pepsin, six fragments are obtained. Of these, four are dipeptides, one is a tripeptide, and one is a hexapeptide. The hexapeptide contains an amino acid which gives a yellow color with ninhydrin, and one of the dipeptides tests positive for sulfur. Give the complete amino acid sequence of peptide F, together with enough of your reasoning to justify your answer.

Problem 5.7

Peptide G has the amino acid composition

(Ala_2,Arg_2,Asp,Glu,Gly,Ile, Leu_2,Met,Phe,Ser,Thr_2,Tyr_2,Val)

Reduction of G with lithium borohydride followed by hydrolysis gives 2-amino-1,3-butanediol.

Tryptic digestion of G gives three products with the compositions:

T1 = (Met,Ser,Thr,Tyr)
T2 = (Ala,Arg,Glu,Ile,Leu,Phe,Val)
T3 = (Ala,Arg,Asp,Gly,Leu,Thr,Tyr)

Digestion of G with chymotrypsin gives four fragments, whose amino acid compositions are:

CT1 = Phe
CT2 = (Met,Thr)
CT3 = (Arg,Asp,Gly,Ser,Tyr)
CT4 = (Ala_2,Arg,Glu,Ile,Leu_2,Thr,Tyr,Val)

Treatment of G with elastase gives five products with the compositions:

E1 = (Ala,Glu,Leu)
E2 = (Ala,Arg,Thr)
E3 = (Ile,Phe,Val)
E4 = (Asp,Gly,Leu,Tyr)
E5 = (Arg,Met,Ser,Thr,Tyr)

Give the complete amino acid sequence of peptide G, together with enough of your reasoning to justify your answer.

Problem 5.8

Peptide H has the amino acid composition

(Arg,Cys,Glu_2,Ile_2,Leu,Lys_2,Met, Phe,Pro_2,Ser,Thr_2,Trp,Tyr)

Treatment of H with lithium borohydride followed by hydrolysis gives 4-amino-5-hydroxypentanoic acid.

Treatment of H with cyanogen bromide gives two products with the compositions:

CB1 = (Cys,Glu,Ile,Leu,Ser,Tyr)
CB2 = (Arg,Glu,Hsr,Ile,Lys_2,Phe,Pro_2,Thr_2,Trp)

Digestion of H with chymotrypsin gives four fragments whose amino acid compositions are:

CT1 = (Ile,Met,Tyr)
CT2 = (Cys,Glu,Leu,Ser)
CT3 = (Glu,Lys,Thr_2,Trp)
CT4 = (Arg,Ile,Lys,Phe,Pro_2)

Tryptic digestion of H gives four products with the compositions:

T1 = (Arg,Ile)
T2 = (Lys,Thr)
T3 = (Glu,Lys,Pro,Thr,Trp)
T4 = (Cys,Glu,Ile,Leu,Met,Phe,Pro,Ser,Tyr)

When H is digested with pepsin, six fragments are obtained. Their amino acid compositions are:

P1 = (Cys,Tyr)
P2 = (Glu,Thr)
P3 = (Lys,Thr)
P4 = (Ile,Met,Phe)
P5 = (Glu,Leu,Ser)
P6 = (Arg,Ile,Lys,Pro_2,Trp)

Give the complete amino acid sequence of peptide H, together with enough of your reasoning to justify your answer.

Problem 5.9

Peptide J has the amino acid composition

(Arg_2,Asp_2,Leu_7,Lys,Met_2,Phe_2, Ser_2,Thr_2,Tyr)

One sample of J is treated with trypsin, and another is treated with chymotrypsin. Aliquots of the two proteolytic digests are mixed, treated with FDNB, and hydrolyzed. When the hydrolyzate is chromatographed on paper, a single yellow spot is found; this is identified as DNP–leucine.

The tryptic digest is resolved into four fragments, whose amino acid compositions are:

T1 = (Arg,Asp,Leu)
T2 = (Arg,Leu,Met,Ser)
T3 = (Leu_2,Met,Thr,Tyr)
T4 = (Asp,Leu_3,Lys,Phe_2,Ser,Thr)

The chymotryptic digest also gives four fragments. Their amino acid compositions are:

CT1 = Leu
CT2 = (Asp,Leu,Phe)
CT3 = (Leu_2,Lys,Met,Thr_2,Tyr)
CT4 = (Arg_2,Asp,Leu_3,Met,Phe,Ser_2)

Cyanogen bromide treatment of peptide J gives three products, two of which are dipeptides. Give the complete amino acid sequence of peptide J, together with enough of your reasoning to justify your answer.

Problem 5.10

The amino acid composition of peptide K is

(Ala_2,Arg_4,Gly_3,Leu_2,Lys_2, Met,Phe,Ser_8,Trp,Tyr)

Treatment of K with cyanogen bromide gives an oligopeptide containing 24 amino acid residues.

Digestion of K with chymotrypsin gives four fragments; their amino acid compositions are:

CT1 = (Arg,Leu,Ser_2)
CT2 = (Arg,Gly,Met,Ser,Tyr)
CT3 = (Ala,Arg,Gly,Lys,Ser_3,Trp)
CT4 = (Ala,Arg,Gly,Leu,Lys,Phe,Ser_2)

When K is treated with pepsin, five fragments are obtained. Their amino acid compositions are:

P1 = Leu
P2 = (Arg,Ser_2,Trp)
P3 = (Arg,Gly,Met,Ser)
P4 = (Ala,Arg,Gly,Lys,Ser_2,Tyr)
P5 = (Ala,Arg,Gly,Lys,Phe,Ser_3)

Digestion of K with trypsin gives seven fragments with the following amino acid compositions:

T1 = (Leu,Ser)
T2 = (Arg,Gly,Met)
T3 = (Lys,Ser_2,Tyr)

T4 = (Arg,Ser_2,Trp)
T5 = (Arg,Leu,Phe,Ser)
T6 = (Ala,Arg,Gly,Ser)
T7 = (Ala,Gly,Ser,Lys)

A mixture of equimolar amounts of fragments T6 and T7 is treated with FDNB and hydrolyzed. When the products are resolved on a paper chromatogram, the only yellow spot found corresponds to DNP–serine. This same fragment mixture is treated with carboxypeptidases A and B to remove their *C*-termini. When the resulting tripeptide mixture is reduced with lithium borohydride and hydrolyzed, the only reduction product obtained is 2-amino-1-propanol.

Give the amino acid sequence of peptide K together with enough of your reasoning to justify your answer.

Problem 5.11

A fragment of human cytochrome c consisting of residues 19 to 54 forms the basis of the following problem. The sequence was originally reported by Matsubara and Smith* and subsequently by Dayhoff.**

The amino acid composition of peptide L is

L= (Ala_3,Arg,Asp_3,Glu_2,Gly_7, His_2,Leu_2,Lys_5,Phe,Pro_2,Ser,Thr_4,Tyr_2,Val)

Reaction of L with FDNB followed by hydrolysis gives DNP-threonine. Reduction of L with lithium borohydride followed by hydrolysis gives 3-amino-4-hydroxybutanoic acid.

Treatment of L with chymotrypsin gives four fragments, with the amino acid compositions:

CT1 = (Ser,Tyr)
CT2 = (Ala_2,Asp_2,Lys,Thr)
CT3 = (Ala,Arg,Glu,Gly_3,Lys,Pro,Thr,Tyr)
CT4 = (Asp,Glu,Gly_4,His_2,Leu_2,Lys_3,Pro,Phe,Thr_2,Val)

Digestion of L with trypsin gives seven fragments. Their amino acid compositions are:

T1 = Lys
T2 = Asp
T3 = (His,Lys)
T4 = (Gly_2,Lys)
T5 = (Glu,Lys,Thr,Val)
T6 = (Arg,Asp,Gly_3,His,Leu_2,Phe,Pro,Thr)
T7 = (Ala_3,Asp,Glu,Gly_2,Lys,Pro,Ser,Thr_2,Tyr_2)

* Matsubara, H. and Smith, E. L., *J. Biol. Chem.*, 238, 2732, 1963.
** Dayhoff, M., *Atlas of Protein Sequence and Structure*, Vol. 5, National Biomedical Research Foundation, 1972, D–13.

When L is treated with pepsin, seven fragments are obtained. Their amino acid compositions are:

P1 = Leu
P2 = (Ser,Tyr)
P3 = (Thr,Val)
P4 = (Gly,His,Leu)
P5 = (Ala_2,Asp_2,Lys,Thr,Tyr)
P6 = (Ala,Arg,Glu,Gly_3,Lys,Phe,Pro,Thr)
P7 = (Asp,Glu,Gly_3,His,Lys_3,Pro,Thr)

Lithium borohydride reduction of P7 followed by hydrolysis gives 3-amino-4-hydroxybutanoic acid, while treatment of P6 in the same way gives 2-aminoethanol.

Digestion of the peptide with elastase gives 12 fragments. Their amino acid compositions are:

E1 = Gly
E2 = Ala
E3 = (Ala,Glu)
E4 = (Thr,Val)
E5 = (Gly,Pro)
E6 = (Gly,Leu,Phe)
E7 = (Asp_2,Lys)
E8 = (Glu,Gly,Lys)
E9 = (Arg,Gly,Lys,Thr)
E10 = (Asp,Gly,His,Leu,Pro)
E11 = (Ala,Ser,Thr,Tyr_2)
E12 = (Gly,His,Lys_2,Thr)

What is the amino acid sequence of peptide L?

Problem 5.12

Treatment of peptide M with 2-mercaptoethanol gives peptide M′, whose amino acid composition is

(Ala_7,Arg_3,Cys_3,Gly,Ile,Lys_2,Met_2,Phe,Ser,Thr, Trp,Tyr,Val)

Treatment of M′ with lithium borohydride followed by hydrolysis gives 2-amino-1,3-propanediol.

Cyanogen bromide treatment of M′ gives three products:

CB1 = (Ala_2,Arg,Cys,Hsr,Phe)
CB2 = (Ala_2,Arg,Cys,Gly,Hsr,Ile,Lys,Trp)
CB3 = (Ala_3,Arg,Cys,Lys,Ser,Thr,Tyr,Val)

Digestion of M′ with chymotrypsin gives four fragments; their amino acid compositions are:

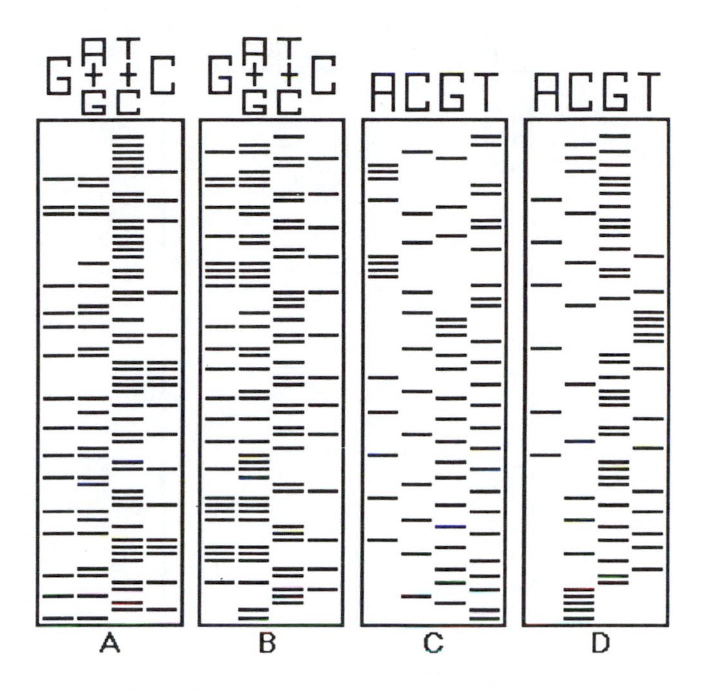

FIGURE 1 Sequencing gels for Problem 5.13.

CT1 = (Ala,Arg,Gly,Ile,Trp)
CT2 = (Ala,Arg,Cys,Ser,Thr,Val)
CT3 = (Ala_2,Cys,Lys,Met,Tyr)
CT4 = (Ala_3,Arg,Cys,Lys,Met,Phe)

Digestion of M′ with trypsin gives six fragments:

T1 = (Ala,Arg,Ile)
T2 = (Ala_2,Arg,Tyr)
T3 = (Ala,Arg,Cys,Met)
T4 = (Ala,Gly,Lys,Trp)
T5 = (Cys,Ser,Thr,Val)
T6 = (Ala_2,Cys,Lys,Met,Phe)

Elastase digestion of M′ yields, among other products, seven dipeptides, one of which contains sulfur.

One of the products of elastase-catalyzed hydrolysis of M is an octapeptide. Give the amino acid sequence of M together with enough of your reasoning to justify your answer.

Problem 5.13

What is the sequence of the DNA strand which served as the template for each of the sequencing experiments depicted in Figure 1?

Problem 5.14

Restriction of a sample of DNA with three different enzymes (Msp I, Abr I, Xor I), both individually and in pairs, gives the gel shown in Figure 2. The standard

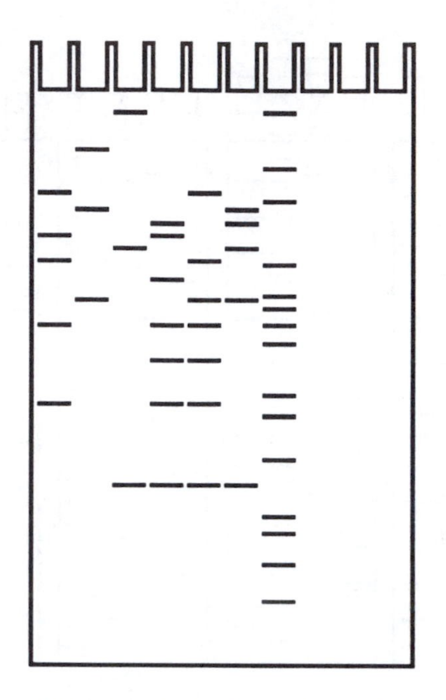

FIGURE 2 Msp I/Abr I/Xor I multiple digest electrophoresis gel (Problem 5.14).

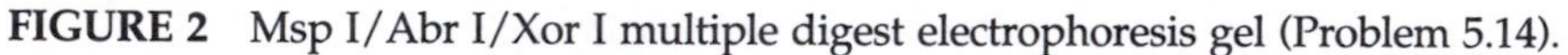

(well 7) is the pGEM DNA Markers, and the identities of the enzymes are listed in Table 1. Deduce the restriction map and predict the fragmentation pattern for a triple digest using Msp I, Abr I, and Xor I on the same DNA.

Table 1

Enzymes Used in Msp I/ Abr I/Xor I Triple Digest (Problem 5.14)

Lane 1	Msp I
Lane 2	Abr I
Lane 3	Xor I
Lane 4	Msp I/Abr I
Lane 5	Msp I/Xor I
Lane 6	Abr I/Xor I

Problem 5.15

This problem will provide data on a number of restriction digest experiments on the DNA of bacteriophage λ, a 48.5 kbp double stranded molecule. By the time you complete the problem you should have a fairly complete restriction map of this important cloning vector.

A. Digestion of λ DNA with Apa I gives fragments of 10.1 and 38.4 kbp, while digestion with Nhe I gives fragments of 13.8 and 34.7 kbp. A Nhe I/Apa I double digest gives fragments of 10.1, 13.8, and 24.6 kbp.

B. Digestion with Xma III gives fragments of 11.8, 16.8, and 19.9 Kbp. A double digest, using Xma III and Nhe I, results in fragments of 2.0, 11.8, 14.8, and 19.9 kbp.

C. Digestion with Eco47 III gives fragments of 11.4, 16.1, and 21.0 kbp. A double digest using Eco47 III and Apa I results in fragments of 10.1, 10.9, 11.4, and 16.1 kbp.

D. When λ DNA is digested with Kpn I, fragments of 1.5, 17.1, and 29.9 kbp are obtained. A Kpn I/Apa I double digest gives fragments of 1.5, 7.0, 10.1, and 29.9 kbp, while a Kpn I/ Eco47 III double digest gives fragments of 1.5, 2.4, 11.4, 16.1, and 17.1 kbp.

E. A Pvu I digest of λ DNA gives fragments of 9.6, 11.9, 12.7, and 14.3 kbp. Redigestion of the Pvu I digest with Eco47 III gives a product mixture containing fragments of 1.3, 5.2, 9.1, 9.6, 11.4, and 11.9 kbp.

F. Restriction of λ DNA with Sma I gives fragments of 8.3, 8.6, 12.2, and 19.4 kbp. A Sma I/Pvu I double digest gives fragments of 4.1, 4.2, 5.4, 6.8, 7.5, 8.6, and 11.9 kbp.

G. Digestion of λ DNA with Sno I gives fragments of 5.3, 5.6, 8.3, 13.1, and 16.2 kbp. A Sno I/Nhe I double digest gives fragments of 5.3, 5.5, 5.6, 7.6, 8.3, and 16.2 kbp, and a double digest using Sno I and Pvu I gives fragments of 0.9, 4.4, 4.4, 5.6, 6.3, 8.3, 8.7, and 9.9 kbp. Show the complete λ restriction map.

Adventures In Sequencing

Biocryptography — A New Science

Three a.m. at the intercept desk deep in the bowels of CIA headquarters was not usually a time of great activity. While the constant chatter of signals traveling to and from distant capitals continued at its normal pace, the quality of the intercepts in the early morning hours was generally not sufficient to justify the presence of a highly regarded cryptanalyst. Rayna Laszlo, recently arrived at Langley from Bismarck, where she had undergone her training, was on duty. Knowing that there probably would not be anything significant in the transmissions, she felt reasonably secure in passing the time studying for her comprehensive examination in Biochemistry, one of the few obstacles remaining between her and the Ph.D. she sought. As fate would have it, had anyone but Rayna been on duty, it is unlikely that there would ever have been a solution to the case that came to be known only by the code name "38 Trip 2 Tire". The name, sounding like a quarterback's signal, seemed appropriate in a town in which the Redskins dominated the news six months out of the year.

Deep into the mysteries of recombinant DNA theory, Rayna had just about lost herself in her studies when a mild commotion roused her from her reverie. Sam, a junior-level clerk who, through his reading of the semitechnical literature, shared some of Rayna's interest in biochemistry, was charging down the hallway, waving a sheet of paper which Rayna recognized as output from the decoding computer. The technology of the day was so highly developed that it was no longer necessary to have individual cryptanalysts pore over every transmission. Instead, the computer, in a matter of minutes, was able to try hundreds of possible coding scenarios, eventually coming up with what Rayna liked to refer to as the "primary transcript."

"Rayna," Sam called breathlessly as he approached her desk. "They've done it. The Russians have solved the sequence of the prenatal elaboration factor."

"My God," she replied. "Now it'll be just a matter of time before they can clone the gene and speed up the process of gestation. They'll be turning out alphas and epsilons at an incredible rate."

"Oh no. It really is *Brave New World*, isn't it?" Sam's literary background often served him well in a branch of government in which reading books did not appear to be a high priority.

"Let me see the sequence," said Rayna.

"Well," Sam said. "It's not really a sequence as such. But they do provide the data for determining it. I don't know enough about sequencing, but I bet you can figure it out. Look at this."

The message, straight from Moscow to the embassy in Washington, is reproduced in Table 1.

Rayna stared briefly at the paper Sam had given her, then set to work reassembling the sequence. Sam understood that the symbols T, CB, CT, and P referred to the products of partial hydrolysis of the peptide by trypsin, cyanogen bromide, chymotrypsin, and pepsin, respectively, but he was not sufficiently versed in their specificities to make much sense of the data. Some time later she

Table 1

Peptide Sequence Data

Composition	(Ala_4,Arg,Asp_4,Glu_5,His,Ile_3,Met_3,Phe,Pro,Ser,Thr_8,Trp_2,Tyr)
T1	(Ala_2,Asp,His,Ile,Met,Thr_2,Trp)
T2	(Ala_2,Arg,Asp_3,Glu_5,Ile_2,Met_2,Phe,Pro,Ser,Thr_6,Trp,Tyr)
CB1	(Asp_2,Glu,Hsr,Ile_2,Thr)
CB2	(Ala_2,Asp,His,Ile,Thr_2,Trp)
CB3	(Ala_2,Arg,Glu_2,Hsr,Phe,Thr_2)
CB4	(Asp,Glu_2,Hsr,Pro,Ser,Thr_3,Trp,Tyr)
CT1	(Thr,Trp)
CT2	(Ala,Arg,Met,Trp)
CT3	(Asp,Glu,Thr,Tyr)
CT4	(Ala_2,Asp,His,Ile,Thr_2)
CT5	(Ala,Asp_2,Glu_4,Ile_2,Met_2,Phe,Pro,Ser,Thr_4)
P1	Glu
P2	Trp
P3	Thr
P4	(Glu,Met)
P5	(Ser,Tyr)
P6	(Asp,Glu,Thr)
P7	(Ala_2,Asp,Thr)
P8	(Ala,Arg,Met,Phe)
P9	(Asp_2,Ile,Thr)
P10	(His,Ile,Thr,Trp)
P11	(Ala,Glu,Thr_2)
P12	(Glue,Ile,Met,Pro,Thr)

flashed that sardonic little grin that Sam always found so engaging, put down her pencil, and showed him the results.

"And that's it?" he cried. "This is almost useless. I may not know as much about protein sequencing as you do, but even I can see that the information isn't sufficient to determine the entire sequence. Either they don't know it all yet or something was lost in transmission."

Rayna just sat back, smiling during Sam's tirade. "Tell me, Sam. What's bothering you about this?"

"Okay. Look, there are 35 amino acid residues here. You've managed to place only half of them in unambiguous positions. You don't know the sequences in positions 3 to 5, 9 and 10, 14 to 16, 20 to 23, 29 to 31, or 33 to 35. What you have is almost useless. How can we beat them in cloning the elaboration factor if we don't know the entire sequence?"

"Relax," she went on. "First of all, I do know the entire sequence. Second, I also know that they don't have the elaboration factor; or if they do, this message has no relation to it."

Sam stared, mystified.

"There are several levels of code in this message," she said. "First, of course, is the Cyrillic substitution code, followed by the English equivalent. That was what the computer gave us. Then the sequence itself is coded in terms of experimental data. But the clue here is that they didn't provide sufficient information to solve the entire sequence. That must mean that there's something else embedded in the message. To find that we have to take the sequence and rewrite it using the standard one-letter code for the amino acids. Here, the code is in this textbook."

Opening the famous blue-covered text to the appropriate page, Rayna showed Sam the list of 22 code letters.

"Strictly speaking there are still ambiguities, but it's easy to resolve them if you assume that the wording has to make sense. Here's the final message. I suppose we should get this to the Director as soon as possible."

Briefly showing her results to Sam, she breezed by, leaving him with his mouth agape, thinking that maybe, just maybe, a Ph.D. in Biochemistry is a good thing to pursue.

Solution

Later that morning, as the graveyard shift ended, Rayna joined Sam for breakfast in the cafeteria. Having consumed orange juice and cereal, they both sat back enjoying coffee, when Sam brought the conversation back to the coded peptide.

"Do me a favor," he began. "Show me how you figured out that message. I still don't get it."

She pushed her coffee cup aside, grabbed a handful of paper napkins to write on, and began her explanation.

"Let's write everything we know at each step. Start with the tryptic cleavage. T2 contains arginine, so it must be the *N*-terminal fragment. That gives us

T2–T1 =
$(Ala_2,Asp_3,Glu_5,Ile_2,Met_2,Phe,Pro,Ser,Thr_6,Trp,Tyr)$–Arg–
$(Ala_2,Asp,His,Ile,Met,Thr_2,Trp)$

"Of the four cyanogen bromide fragments, CB2 must be from the original *C*-terminus since it lacks homoserine. The composition of CB2 is identical to that of T1, except for the presence of methionine in T1. This methionine then must be at the *N*-terminus of T1, giving

T2–T1 =
$(Ala_2,Asp_3,Glu_5,Ile_2,Met_2,Phe,Pro,Ser,Thr_6,Trp,Tyr)$–Arg–Met–
$(Ala_2,Asp,His,Ile,Thr_2,Trp)$

"There is only one arginine present in the original peptide; this is found in fragment CB3. The sequence of CB3 is (Ala_2,Glu_2,Phe,Thr_2)–Arg–Met. Since CB3 and CB2 make up the *C*-terminal region of the peptide, only CB1 and CB4 remain to be located. (We can show Figure 1.)

"Chymotryptic fragment CT4 comes from the original *C*-terminus. Except for the lack of tryptophan, CT4 is identical to the unknown *C*-terminal region, CB2. So CB2 can be shown as Trp–$(Ala_2,Asp,His,Ile,Thr_2)$. The *C*-terminal sequence (residues 19 to 35) is

CB3–CB2 = (Ala_2,Glu_2,Phe,Thr_2)–Arg–Met–Trp–$(Ala_2,Asp,His,Ile,Thr_2)$

"There are only two tryptophan residues in the original peptide. One of them is in the sequence Thr–Trp in CT1, and the other one is in CT2, which also contains the only arginine. The *C*-terminal region of CT2 has to be Arg–Met–Trp. Since three of the four residues of CT2 are known, the fourth goes at the *N*-terminus, and CT2 = Ala–Arg–Met–Trp. We already saw that this is the same tryptophan found at the *N*-terminus of CB2. The *C*-terminal sequence of the peptide is CT2–CT4, so the entire peptide can be shown as Figure 2.

"Of the four aromatic amino acids, one tryptophan (in CB2) comes after phenylalanine. But the remaining tryptophan and the single tyrosine are both found in CB4 which, as we have already seen, must come before the *C*-terminal sequence CB3–CB2. Consequently we can say that CT5, which contains phenylalanine, must follow both CT1 and CT3 and must come just prior to CT2, giving Figure 3.

"The tryptophan of CT1 and the tyrosine of CT3 both precede the two methionines of CT5. If we go back to the cyanogen bromide sequence, the only way this can happen is for CB4 to come before CB1. Combining that conclusion with what we have just seen for the chymotryptic sequence gives us Figure 4."

"Hold on a minute," Sam interrupted. "You just lost me. It's been a long night, and I need some more coffee. How about a refill?"

"Great," Rayna replied. "Mine must be pretty cold by now."

As Sam went in search of more coffee, Rayna collected the napkins she had strewn about the table. Someday I've got to get a little more organized, she thought.

(CB1,CB4)-CB3-CB2 =

$$\left[\begin{array}{c}(Asp_2,Glu,Ile_2,Thr)\text{-}Met\\(Asp,Glu_2,Pro,Ser,Thr_3,Trp,Tyr)\text{-}Met\end{array}\right]\text{-}(Ala_2,Glu_2,Phe,Thr_2)\text{-}Arg\text{-}Met\text{-}(Ala_2,Asp,His,Ile,Thr_2,Trp)$$

FIGURE 1 Partial cyanogen bromide sequence of the unknown peptide.

(CT1,CT3,CT5)-CT2-CT4 =

$$\left[\begin{array}{c}Thr\text{-}Trp\\(Asp,Glu,Thr)\text{-}Tyr\\(Ala,Asp_2,Glu_4,Ile_2,Met_2,Pro,Ser,Thr_4)\text{-}Phe\end{array}\right]\text{-}Ala\text{-}Arg\text{-}Met\text{-}Trp\text{-}(Ala_2,Asp,His,Ile,Thr_2)$$

FIGURE 2 Partial chymotryptic sequence of the unknown peptide.

(CT1,CT3)-CT5-CT2-CT4 =

[Thr-Trp / (Asp,Glu,Thr)-Tyr]-(Ala,Asp$_2$,Glu$_4$,Ile$_2$,Met$_2$,Pro,Ser,Thr$_4$)-Phe-Ala-Arg-Met-Trp-(Ala$_2$,Asp,His,Ile,Thr$_2$)

FIGURE 3 Partial chymotryptic sequence of the unknown peptide.

(CT1,CT3)-CT5-CT2-CT4 =

[Thr-Trp / (Asp,Glu,Thr)-Tyr]-(Glu,Pro,Ser,Thr)-Met-(Asp$_2$,Glu,Ile$_2$,Thr)-Met-(Ala,Glu$_2$,Thr$_2$)-

Phe-Ala-Arg-Met-Trp-(Ala$_2$,Asp,His,Ile,Thr$_2$)

FIGURE 4 Partial chymotryptic sequence of the unknown peptide.

When Sam returned, she began again.

"Residues 1 to 11 are contained in CB4. But the first six residues are also in CT1 and CT3."

"I have it," Sam burst in. "To get positions 7 to 11 you just subtracted the contents of CT1 and CT3 from CB4, and put in what was left. Now, what's next?"

"The peptic digestion products include one fragment, P3, which contains only threonine. This can arise only from the *N*-terminus of the original peptide. The tryptophan in fragment P2 is at position 2 so that the *N*-terminal hexapeptide is Thr–Trp–(Asp,Glu,Thr)–Tyr

"Fragment P6 contains aspartate, glutamate, and threonine, corresponding to positions 3 to 5. Since this fragment contains both aspartate and glutamate, we must conclude that one of them is in the amide form, that is asparagine or glutamine. We can use the abbreviations Asx and Glx to indicate uncertainty. Either aspartate or glutamate may be at the *N*-terminus of P6; there is no information on the *C*-terminus, so this fragment remains ambiguous. The sequence as we know it so far is Thr–Trp–(Asx,Glx,Thr)–Tyr–(Glu,Pro,Ser,Thr)–Met–(Asp_2,Glu,Ile_2,Thr)–Met–(Ala,Glu_2,Thr_2)–Phe–Ala–Arg–Met–Trp–(Ala_2,Asp,His,Ile,Thr_2)

"The tyrosine residue at position 6 is the only tyrosine in the peptide. Fragment P5 (Tyr–Ser) contains this residue, and must follow P3, P2, and P6:

P3–P2–P6–P5 = Thr–Trp–(Asx,Glx,Thr)–Tyr–Ser

"The next four positions, 8 to 11, contain the sequence (Glu,Pro,Thr)–Met; these residues are found in fragment P12, whose sequence must be Glu–(Pro,Thr)–Met–Ile. Again, no data are provided to assign the (Pro,Thr) sequence. The *N*-terminal region, positions 1 to 12, is

P3–P2–P6–P5–P12 = Thr–Trp–(Asx,Glx,Thr)–Tyr–Ser–Glu–(Pro,Thr)–Met–Ile

"Now let's examine the *C*-terminal region. The tryptophan residue at position 28 is contained in fragment P10; positions 28 to 31 are in the sequence Trp–(Ile,His,Thr). The *C*-terminal tetrapeptide (Ala_2,Asp,Thr) is found in fragment P7. Positions 28 to 35 can be shown as

P10–P7 = Trp–(His,Ile,Thr)–Asp–(Ala_2,Thr)

Immediately preceding this grouping is the sequence Phe–Ala–Arg–Met, which is found in fragment P8. The *C*-terminal region can now be shown as

P8–P10–P7 = Phe–Ala–Arg–Met–Trp–(His,Ile,Thr)–Asp–(Ala_2,Thr)

"At this point the sequence is that shown in Figure 5. The only methionine-containing peptic fragment which has not yet been located is P4, which contains the methionine at position 18. The sequence of P4 is Glu–Met. Immediately preceding P4 and immediately following P12 is the grouping (Asp_2,Ile,Thr), which corresponds to fragment P9. The *N*-terminal amino acid of P9 must be

P3-P2-P6-P5-P12-(P1,P4,P9,P11)-P8-P10-P7 =

Thr-Trp-(Asx,Glx,Thr)-Tyr-
Ser-Glu-(Pro,Thr)-Met-Ile-

[Glu
(Asp_2,Ile,Thr)
(Glu,Met)
$Glu-(Ala,Thr_2)$]

-Phe-Ala-Arg-Met-Trp-
$(Ile,His,Thr)-Asp-(Ala_2,Thr)$

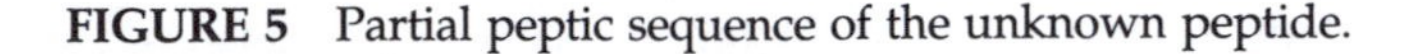

FIGURE 5 Partial peptic sequence of the unknown peptide.

aspartate, while the other aspartate residue must have arisen from asparagine. Thus P9 = Asp–(Asn,Ile,Thr).

"The grouping (Ala,Glu_2,Thr_2) follows P4. The only two peptic fragments remaining are P1 (Glu) and P11 (Ala,Glu,Thr_2). It is clear that P11 = Glu–(Ala,Thr_2), but the order of P1 and P11 is uncertain. The two possibilities are

$$P1–P11 = Glu–Glu–(Ala,Thr_2)$$

and

$$P11–P1 = Glu–(Ala,Thr_2)–Glu$$

So position 19 contains glutamate, but we don't know whether the glutamate is contained in fragment P1 or fragment P11.

"Before we write the entire sequence again, let's go back to P6. We showed it as (Asx,Glx,Thr), but we can add a refinement. Either aspartate or glutamate must be at the *N*-terminus. As a result P6 = Asp–(Gln,Thr) or Glu–(Asn,Thr).

"Now the collection of peptic fragments can be shown as in Figure 6.

"As you pointed out last night, Sam, there is still considerable ambiguity in the sequence. But look at what develops when the one-letter amino acid codes are used (Figure 7).

"Position 3 contains either D or E, and positions 4 and 5 will contain either Q or *N* at one position and T at the other. Since it's unlikely that any word beginning with TW will not be followed by a vowel, it's reasonable to assume that the third position will be E (glutamate). Only two possible combinations remain, namely TWENT or TWETN. If we include the known Y at position 6, a brief inspection should convince you that the word in question is TWENTY.

"The next four positions can be SEPT or SETP. SEPT would be in keeping with the assignment of a date (20 September). Positions 11 to 17 establish the time of day (MIDNITE).

"In the remainder of the message the word FARM stands out, followed by what appears to be WITH. The last word can be found by simply scrambling the letters D(AAT) to come up with DATA.

"In the beginning of this section the sequence is either MEE(ATT) or ME(ATT)E. While the word MEAT seems to pop up, the TE that would follow makes no sense. Only the first possibility seems to form two words, namely MEET AT. This fits nicely with the following word, FARM. The whole message

FIGURE 6 Partial peptic sequence of the unknown peptide.

TW [D(QT) / E(NT)] YSE(PT)MID(INT)EM [E / E(ATT)] FARMW(IHT)D(AAT)

FIGURE 7 Sequence of unknown peptide using one-letter amino acid code.

is: TWENTY SEPT, MIDNITE. MEET AT FARM WITH DATA. And the unambiguous amino acid sequence is Thr–Trp–Glu–Asn–Thr–Tyr–Ser–Glu–Pro–Thr–Met–Ile–Asp–Asn–Ile–Thr–Glu–Met–Glu–Glu–Thr–Ala–Thr–Phe–Ala–Arg–Met–Trp–Ile–Thr–His–Asp–Ala–Thr–Ala."

Sam stared at the last of Rayna's paper napkins, raised his cup as in a toast, and slowly sipped his coffee.

"I'm impressed," he said. "Has anyone ever used biological macromolecules to encode messages before?"

"Not to my knowledge," she replied. "It appears that someone over there has invented a whole new class of cryptographic methods. Before we came to breakfast I dropped a memo to the Director, explaining, in simplified terms of course, what we found. I also suggested that in the future we refer to this technique as "Biocryptography". It sounds good, and I think the word pretty well describes the concept."

"I agree," Sam said. "You know, after you get your Ph.D. you might want to use some of this stuff in your teaching. You might even write a book about it."

"Don't think I haven't considered that already, Sam."

Rayna checked her watch, suddenly feeling the weight of a momentous night on her shoulders and eyelids.

"Sam," she concluded. "I'm wiped. I've got to get home and hit the sack. See you tonight."

As he watched her leave, Sam once again raised his cup, tilted it toward her and thought: Here's looking at you kid. Almost immediately, with a knitted brow and a wry grin he whispered to himself, "Did I really say that?"

A Night at the Opera (House)

Despite the normally slow progress of ideas in the federal bureaucracy, it didn't take long for Rayna's superiors at the CIA to recognize her talent and to reward her for solving the "38 Trip 2 Tire" case. Although she received a merit salary raise and a substantial bonus, the real mark of upward mobility at the agency was her move to a new office which, aside from being marginally larger, had that most coveted of all perquisites, a window. As any federal bureaucrat knows, this is the truest indicator of one's standing in the agency. Rayna's future looked bright.

Sam, in the meantime, had become bolder in professing his affection for Rayna, and while stopping short of any overt displays, he did on a few occasions invite her to a movie, invariably a bargain matinee, and dinner at a modest local restaurant. It was hardly surprising, then, that Rayna chose to celebrate her "promotion" with Sam, inviting him to accompany her to dinner followed by a performance of one of the great Russian ballet companies in the opera house of the John F. Kennedy Center for the Performing Arts. This was to be a gala benefit affair, with the Friends of the Kennedy Center reaping the profit on tickets which sold for a hefty premium over their box office price. The occasion was made even more significant by the anticipated attendance of the U.S. and Russian presidents. The Presidential box in the first tier would be decorated in bunting, and the entire staff of ushers would undoubtedly be thrown into a tizzy as Secret Service agents scoured the hall prior to the arrival of the dignitaries.

On the appointed evening, Rayna's roommate held the door ajar, allowing Sam to enter the small but comfortable apartment. This was not Sam's first visit, of course, so it was with only mild surprise that he took note of the coffee table on which were spread a text on restriction analysis of DNA and a collection of small photographs. Good old Rayna, he thought, never wasting a minute, constantly boning up on her biochemistry. Sam recognized that the photographs depicted agarose gels showing the locations of various electrophoretically separated DNA fragments. He thumbed casually through the book, failing to notice as she entered the room. When he finally did look up he was pleased to see her already standing there.

Sam's disappointment over the rather pretentious and vaguely unappetizing dinner at the fanciest of the Kennedy Center's restaurants was mitigated by Rayna herself, whom he found, as usual, utterly charming, despite her tendency to lecture him on the finer points of her latest professional passion, restriction analysis. Just as Sam began to suggest that Rayna not mix business with pleasure, the obsequious head-waiter leaned over her shoulder, and whispered into her ear.

"I'm afraid we're going to have to miss the precurtain cocktail party, Sam," she said. "It looks like something's come in at the company, and I have to get over there. Would you mind?"

Having little choice, Sam trailed behind as Rayna left the restaurant, making her way downstairs and through the flag-draped Hall of the States, exiting the building and quickly entering a waiting agency car. During the short ride to Langley, Sam occasionally stole a surreptitious glance at her, but realized that she would brook none of his small talk. Rayna, although none too pleased to have her evening interrupted, managed to avoid complaining.

Once in her office she began examining the data which had been left for her. Sam, looking over her shoulder, again recognized the pictures of DNA gels. He retrieved a sheet of paper containing a series of abbreviations which he knew were labels denoting the restriction enzymes responsible for each of the gel patterns.

"I don't get it," Sam said. "What's the big deal? I mean, here is a set of restriction experiments. There's no big secret here. I think even I could figure out the restriction map for this DNA."

"That's exactly the point, Sam," Rayna replied. "Why is all this stuff on my desk? It's essentially trivial. Why would anyone want to send this stuff anyway? These are the same gels you were looking at back at the apartment. I've been studying the pictures for a few days; the rightmost lane in each gel must be a

Table 2

Identities of Restriction Enzymes for Gels 1 to 4

Gel 1	Gel 2	Gel 3	Gel 4
Aat I	Kpn I	Aat I/Mro I	EcoR V/Kpn I
Avr I	Mla I	Aat I/Sac I	EcoR V/Sac I
BamK I	Mro I	Avr I/Nar I	Gsb I/Sac I
Eco47 III	Nar I	Avr I/Sac I	Hind III/Mla I
EcoR V	Sac I	Avr I/Spc I	Mla I/Sac I
Gsb I	Sna I	Bamk I/Hind III	Mro I/Sna I
Hind III	Spc I	Eco47 III/Gsb I	Sna I/Spc I
		Eco47 III/Sac I	

molecular weight standard. From its appearance I'd guess that it's part of the commercial 1 Kb Ladder. Normally that shows about 22 bands, but there are only 15 here. The pattern seems to coincide with the smaller fragments — the ones from about 0.7 kb up to about 5 kb. I've already measured all the migration distances and plotted the graph of distance as a function of the log of the size for the standards."

Flipping on her desktop computer, Rayna entered a few commands, finally bringing a spreadsheet up on the screen. After a few more keystrokes her printer jumped to life.

"I used a regression template to calculate the slope and intercept of the standard curve, and then I calculated the sizes of the other fragments from their migration distances. This printout summarizes all the data. But it all made no sense without the list of enzymes. Why don't you see if you can solve the restriction map? I've got to think this thing through."

The list of enzymes is displayed in Table 2, and the gels were as shown in Figure 8.

Using Rayna's calculations, Sam set about locating each of the restriction sites on the DNA molecule. Some time later he looked up.

"That wasn't too difficult," he said. "Although I found the Eco47 III/Gsb I double digest data confusing. It took a while to realize that the 1.8 kbp fragment in the double digest was identical to the one found in the Eco47 III digest, rather than in the Gsb I digest. The assignments I made based on the Eco47 III/Gsb I double digest were confirmed by the Eco47 III/Sac I double digest experiment. So here it is, the entire restriction map. Now what?"

Rayna mirrored the quizzical look on Sam's face, until she suddenly lurched to the wall to study a multicolored chart listing some 400 restriction enzymes and their recognition sequences. As she consulted a textbook and started taking notes, her lips began to quiver. Finally she whirled to face Sam, her body shaking.

"Sam," she whispered urgently. "Do whatever you have to, but get us a car and a driver who can get us back to the Kennedy Center as fast as possible."

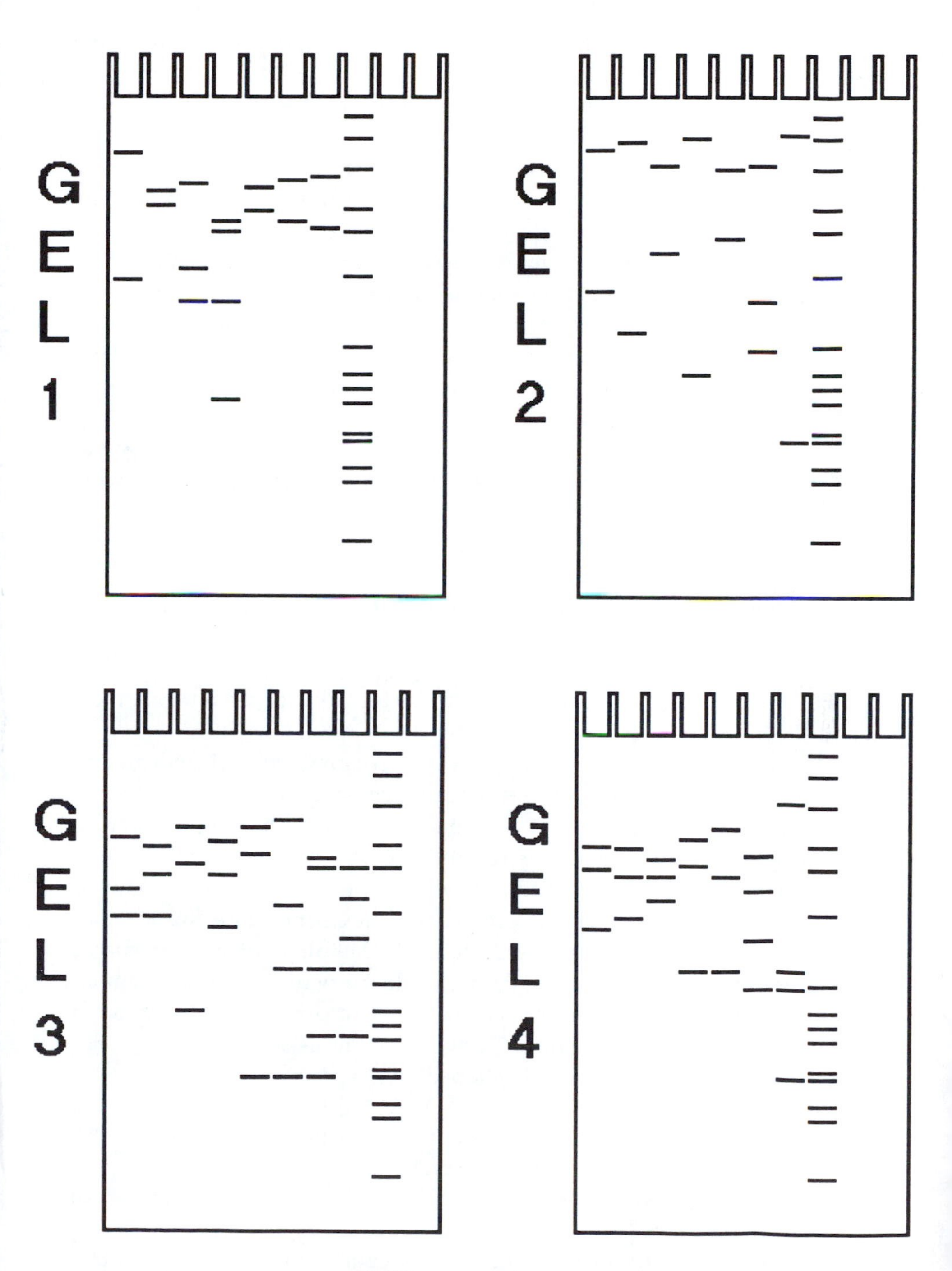

FIGURE 8 Gels 1 to 4.

He looked at her inquisitively. "I'll explain in the car," she said. "But do it. I know this sounds awfully melodramatic, but, believe it or not, the fate of the world is hanging in the balance."

Sam almost laughed. This is like something out of a bad movie, he thought. But Rayna's drawn face convinced him that the crisis was real.

As they sped through the night, Rayna used the mobile car phone to call the Kennedy Center house manager, who was overseeing the cocktail party. She whispered into the mouthpiece, finally convincing him to meet them at one of the non-public entrances to the massive granite building. Sam overheard her requesting information about rehearsal rooms and (of all things) urns, all the while trying to avoid any panic at either end of the conversation.

The call completed, she turned to Sam. "It's not the quantitative restriction map that's important, Sam," she said. "That's the beginning, but all we really need is the sequence of restriction sites."

"We know the sequence," Sam replied. "I've been trying to piece together the names of the enzymes, arranging them in various permutations and combinations, but nothing makes sense."

"The names of the enzymes aren't important. It's the recognition sequences that contain the message."

"But how can that be?" he asked. "The alphabet of the sequences is just the same as that of DNA. There are only four letters."

"Remember the codon dictionary..."

Before Rayna could explain further, the car turned a corner, bringing them to a well-hidden loading dock.

"We're here. I'll finish the explanation later. For now, just stick with me and back me up. There may be some resistance to what we have to do."

Just as they drove up, the house manager appeared, making no effort to conceal his displeasure at being interrupted during a presidential appearance. Rayna identified herself and asked to be directed to Rehearsal Room Five. A small elevator whisked them to the second tier of the Opera House. Rushing past the women's restroom they quickly descended a half-flight of stairs, crossed a red-carpeted vestibule, and passed through a doorway that was neatly camouflaged in the mirrored wall. A short hallway led to a nondescript door bearing a small red sign inscribed "Studio 5." Once there, she asked the manager once again to describe the urn he had mentioned on the phone.

"It's the most peculiar thing," he said. "One of the little ballerinas discovered it the other day, but no one could move it. It seems to be bolted to the floor. No one was able to twist the cover off it, either."

"I know this sounds strange," said Rayna. "But has anyone found a small glass statue of a moose?"

"My goodness," the manager cried. "How could you have known that? It was sitting behind the curtain when the maintenance man found it. He gave it to the same ballerina, as a gift."

"Get it quickly," ordered Rayna. "Then leave us."

Sam stared open-mouthed, wanting to ask if Rayna had finally gone off the deep end. She held up a hand indicating that no questions would be tolerated, and proceeded to extract what looked like a small bottle opener from her purse. Running it gingerly around the top of the mysterious urn, she finally managed to force the cover off, by which time the manager had returned with a glass animal barely the size of a tennis ball. Rayna took the statuette, and with a sharp glance, dismissed the manager. She gently placed the statue into the urn, manipulated it until it locked in place, then stood up and started searching the walls.

"Help me, Sam," she implored. "There's a utility duct running through here somewhere. Time's running out and we have to find it."

Desperately they ran their hands over the walls, across the floor, into the corners of the room, until Sam, displacing a piece of stage scenery, discovered a small door built into the wall. Behind the door Rayna found a small wheel.

"The gas valve! Sam, please turn it and shut it down."

As Sam strained to close the valve, Rayna again reached into the urn, grabbed the glass statue, twisted it until its legs broke off, and quickly removed it. She showed it to Sam, pointing out the unusual coloration which it began to assume.

"It's a binary, Sam. Whatever was in here is harmless until it comes in contact with one of the components of ordinary cooking gas. The result is lethal."

Sam blanched. "My God! You mean someone was trying to release a poison gas in here. But why?"

"Not here in the practice room. The binary would have entered the line and passed into the Opera House."

"Oh no!" The two presidents! Good Lord, the world would have been plunged into…But how did you know?"

"Later, Sam," Rayna smiled. "Right now I'm in the mood for a little Balanchine and Stravinsky. Let's get to our seats."

Fortunately the walk to their seats was not long, since the only tickets available to Rayna had placed them in the last row of the second tier, about as far as possible from the stage. As they took their seats, she whispered, "By the way, did you notice anything peculiar about the house manager?"

"No," replied Sam. "Why?"

"I don't know. Something about him reminded me of *The Phantom of the Opera*, but I can't quite place it."

Sam turned to her, wide-eyed. But before he could say anything, the house lights dimmed and the orchestra began a raucous, if somewhat Russified rendition of "Hail to the Chief."

Solution

The performance by the Russian troupe was, as expected, technically flawless. Nonetheless, as they joined the crowd leaving the opera house and heading for public transportation, Rayna voiced her disappointment to Sam.

"You know, they needn't have filled the program with the staid old classics. Their repertory has a couple of contemporary experimental works that are supposed to be real stunners."

"Yes," replied Sam. "But sometimes you have to play to the audience. I don't think that either of the presidents would have appreciated anything so shocking.

"And speaking of shocking," he continued, unfolding the computer printout she had given him earlier, "would you mind explaining to me that business with the moose and the jug?"

"Actually, it was an elk and a vase," she chuckled. "Quite an important difference. Let's go back to the gels. The printout (Table 3) shows the slope and intercept calculated from the standard 1 Kb Ladder fragments in each picture. Below that are the migratory distances in millimeters and the calculated restriction fragment lengths.

"It was immediately clear that we were looking at a DNA chain of 4.5 kbp. As you said, constructing the restriction map wasn't too difficult."

Sam's map, which Rayna extracted from her purse, is shown in Figure 9.

Table 3

Calculated Sizes of Restriction Fragments (Gels 1 to 4)[a]

	Gel 1		Gel 2		Gel 3		Gel 4	
	Distance*	kb	Distance	kb	Distance	kb	Distance	kb
Lane 1	**Aat I**		**Kpn I**		**Aat I/Mro I**		**EcoR V/Kpn I**	
	8.2	3.5	7.7	3.6	15.2	2.2	16.3	2.0
	26.6	1.0	28.0	0.9	22.7	1.3	19.5	1.6
					26.8	1.0	28.0	0.9
	Sum:	4.5		4.5		4.5		4.5
Lane 2	**Avr I**		**Mla I**		**Aat I/Sac I**		**EcoR V/Sac I**	
	13.6	2.4	6.5	3.9	16.4	2.0	16.2	2.0
	15.8	2.1	34.1	0.6	20.7	1.5	20.4	1.5
					26.8	1.0	26.4	1.0
	Sum:	4.5		4.5		4.5		4.5
Lane 3	**BamK I**		**Mro I**		**Avr I/Nar I**		**Gsb I/Sac I**	
	12.5	2.6	9.7	3.2	13.8	2.4	17.9	1.8
	25.1	1.1	22.5	1.3	19.3	1.7	20.6	1.5
	29.8	0.8			40.4	0.4	24.0	1.2
	Sum:	4.5		4.5		4.5		4.5
Lane 4	**Eco47 III**		**Nar I**		**Avr I/Sac I**		**Hind III/Mla I**	
	17.9	1.8	5.9	4.1	15.8	2.1	15.2	2.2
	19.6	1.6	40.1	0.4	20.8	1.5	19.1	1.7
	29.8	0.8			28.3	0.9	34.2	0.6
	44.0	0.3						
	Sum:	4.5		4.5		4.5		4.5

Table 3 (continued)

Calculated Sizes of Restriction Fragments (Gels 1 to 4)[a]

	Gel 1		Gel 2		Gel 3		Gel 4	
	Distance*	kb	Distance	kb	Distance	kb	Distance	kb
Lane 5	**EcoR V**		**Sac I**		**Avr I/Spc I**		**Mla I/Sac I**	
	13.1	2.5	10.3	3.0	13.8	2.4	13.6	2.4
	16.3	2.0	20.4	1.5	17.7	1.9	20.6	1.5
					50.1	0.2	34.2	0.6
	Sum:	4.5		4.5		4.5		4.5
Lane 6	**Gsb I**		**Sna I**		**BamK I/Hind III**		**Mro I/Sna I**	
	12.0	2.7	9.8	3.2	12.5	2.6	17.3	1.9
	17.9	1.8	29.8	0.8	25.2	1.1	22.5	1.3
			36.8	0.5	34.3	0.6	29.8	0.8
					50.0	0.2	36.8	0.5
	Sum:	4.5		4.5		4.5		4.5
Lane 7	**Hind III**		**Spc I**		**Eco47 III/Gsb I**		**Sna I/Spc I**	
	11.4	2.8	5.4	4.3	18.0	1.8	9.8	3.2
	19.0	1.7	49.7	0.2	19.6	1.6	34.2	0.6
					34.4	0.6	36.9	0.5
					44.0	0.3	50.0	0.2
					50.0	0.2		
	Sum:	4.5		4.5		4.5		4.5
Lane 8					**Eco47 III/Sac I**			
					19.7	1.6		
					24.1	1.2		
					30.0	0.8		
					34.4	0.6		
					44.1	0.3		
	Sum:					4.5		

[a] Slope = –33.7, intercept = 128 (for each gel).
* All distances are in millimeters.

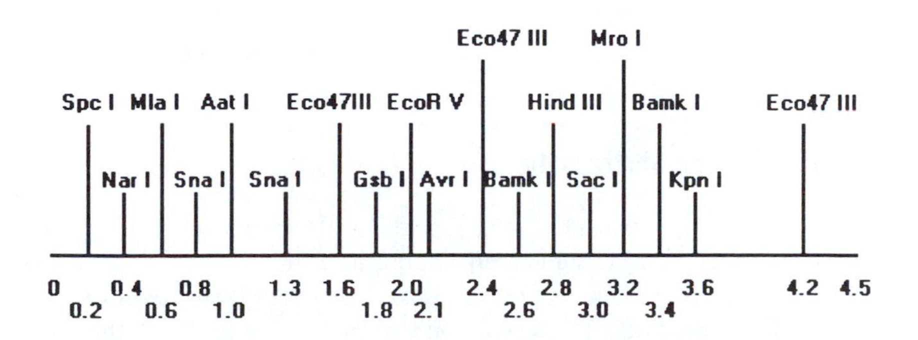

FIGURE 9 Restriction map for Kennedy Center gels.

"I realized," she continued, "that there could be no special significance to the actual names of the enzymes. You recognized that, too. Even looking at the base sequences of the recognition sites was no help. As you pointed out, the four-letter alphabet of DNA doesn't allow for much direct information. But there was a clue in the fact that each of the enzymes has a six base pair recognition site (Table 4). This immediately suggested that the recognition sites could be seen as codons, each representing a pair of amino acids.

"I transcribed the codons into their RNA complements and then translated those into one letter amino acid symbols (Table 5). Of course the coding was fairly simple-minded in that, while each of the codons was transcribed as its antiparallel complement, the entire chain was read from left to right. Consequently, although each triplet is transcribed from 3′ to 5′, transcription begins at the 5′-end of the DNA chain. It was only when I realized this that the final message jumped out at me."

Rayna showed Sam yet another scrap of paper she had hidden in her purse. The message was STAGEFYVPRYVASELIDGLASSELKLEGSSETGAS.

"The only ambiguity was in the codons for Avr I. Four possible letter combinations are possible, but as you'll see, the only one that makes any sense is 'GL'."

Puzzled, Sam stared at the message.

"Well," he began, "the first word is obviously STAGE, and the last is GAS. Just before that must be SET. That leaves FYVPRYVASELIDGLASSELKLEGS. Oh, I see! If you continue to work from right to left things begin to pop out. LEGS, ELK, GLASS, LID, VASE, PRY. But what's FYV?"

"Allow for a little poetic license," Rayna said coyly. "Pronounce it as it's written."

"Oh," Sam replied, slightly shamefaced. "FIVE. So the whole message is 'Stage five; pry vase lid; glass elk legs set gas'."

"From there it wasn't too hard to figure out where and why things were scheduled to happen," she concluded. "With the help of the house manager's information, everything fell neatly into place."

As they approached Rayna's apartment Sam reflected on the events of the preceding few hours. She's really something, he thought.

Rayna opened the door, turned to Sam, and asked cheerfully, "Care to join me for a nightcap, Sam?"

From Baltimore with Love

A Middle Eastern open-air market, transmogrified and transported to Yuppieville, U.S.A. Harborplace. Rayna could barely suppress a grin as she and Sam meandered through the long pavilion bordering the waterfront. The crown jewel of Baltimore's downtown urban renewal, the two building complex at the Inner Harbor reflected both the best and the worst of modern shopping. A hundred yards away, making a right angle with the first building, the second structure

Table 4

Recognition Sequences for Selected Restriction Endonucleases[a]

Aac I	GGATCC	Cpe I	TGATCA	Mro I	TCCGGA
Aae I	GGATCC	Csp I	CGGACCG	Msp I	CCGG
Aat I	AGGCCT	Csp I	CGGTCCG	Nae I	GCCGGC
Aat II	GACGTC	Csp45 I	TTCGAA	Nar I	GGCGCC
Abr I	CTCGAG	Dds I	GGATCC	Nbl I	CGATCG
Acy I	GACGCC	Dra I	TTTAAA	Nco I	CCATGG
Acy I	GACGTC	Ecl XI	CGGCCG	Nde I	CATATG
Acy I	GGCGCC	Eco47 III	AGCGCT	Nhe I	GCTAGC
Acy I	GGCGTC	EcoICR I	GAGCTC	Not I	GCGGCCGC
Afl II	CTTAAG	EcoR I	GAATTC	Nru I	TCGCGA
Ali I	GGATCC	EcoR V	GATATC	Nsi I	ATGCAT
Alu I	AGCT	Fsp I	TGCGCA	Nsp I	ACATGC
Alw44 I	GTGCAC	Gdi I	AGGCCT	Nsp I	ACATGT
Ama I	TCGCGA	Gdi II	CGGCCG	Nsp I	GCATGC
Apa I	GGGCCC	Gdi II	TGGCCG	Nsp I	GCATGT
Ase I	ATTAAT	Gdo I	GGATCC	Pst I	CTGCAG
Asn I	ATTAAT	Gin I	GGATCC	Pvu I	CGATCG
AspH I	GAGCAC	Gox I	GGATCC	Pvu II	CAGCTG
AspH I	GAGCTC	Gsb I	CTCCAG	Rhe I	GTCGAC
AspH I	GTGCAC	Hae I	AGGCCA	Rhs I	GGATCC
AspH I	GTGCTC	Hae I	AGGCCT	Rsh I	CGATCG
AtuC I	TGATCA	Hae I	TGGCCA	Rsp I	CGATCG
Ava I	CCCGAG	Hae I	TGGCCT	Rsr I	GAATTC
Ava I	CTCGAG	Hae II	AGCGCC	Sac I	GAGCTC
Ava I	CCCGGG	Hae II	AGCGCT	Sac II	CCGCGG
Ava I	CTCGGG	Hae II	GGCGCC	Sal I	GTCGAC
Ava III	ATGCAT	Hae II	GGCGCT	Sca I	AGTACT
Avr I	CCCGAG	HgiA I	GAGCAC	Sfu I	TTCGAA
Avr I	CCCGGG	Hinb III	AAGCTT	Sma I	CCCGGG
Avr I	CTCGAG	Hinc II	GTCAAC	Sna I	GTATAC
Avr I	CTCGGG	Hinc II	GTCGAC	SnaB I	TACGTA
Avr II	CCTAGG	Hinc II	GTTAAC	Sno I	GTGCAC
Bac I	CCGCGG	Hinc II	GTTGAC	Spa I	GCATGC
Bal I	TGGCCA	Hind III	AAGCTT	Spc I	ACTAGT
BamF I	GGATCC	Hinf I	GAATC	Sph I	GCATGC
BamH I	GGATCC	Hinf I	GACTC	Ssp I	AATATT
BamK I	GGATTC	Hinf I	GAGTC	Stu I	AGGCCT
BamN I	GGATCC	Hinf I	GATTC	Xba I	TCTAGA
Bbe I	GGCGCC	Hpa I	GTTAAC	Xho I	CTCGAG
Bbr I	AAGCTT	Hsu I	AAGCTT	Xho II	AGATCC
BbrP I	CACGTG	Kpn I	GGTACC	Xho II	AGATCT
Bbu I	GCATGC	Ksp I	CCGCGG	Xho II	GGATCC
Bcl I	TGATCA	Mbo I	AGATC	Xho II	GGATCT
Bgl II	AGATCT	Mbo I	CGATC	Xma I	CCCGGG
Bpe I	AAGCTT	Mbo I	GGATC	Xma III	CGGCCG
BseP I	GCGCGC	Mbo I	TGATC	Xni I	CGATCG
Bst I	GGATCC	Mki I	AAGCTT	Xor I	CTGCAG
Cla I	ATCGAT	Mla I	TTCGAA	Xor II	CGATCG

[a] Source: *Recognition Sequences of Restriction Endonucleases and Methylases,* Boehringer Mannheim BmbH, Biochemica, 1985.

Table 5

The Genetic Code Word Dictionary[a]

1st	2nd U	C	A	G	3rd
	UUU Phe	UCU Ser	UAU Tyr	UGU Cys	U
	UUC Phe	UCC Ser	UAC Tyr	UGC Cys	C
U	UUA Leu	UCA Ser	UAA Ochre	UGA Umber	A
	UUG Leu	UCG Ser	UAG Amber	UGG Trp	G
	CUU Leu	CCU Pro	CAU His	CGU Arg	U
	CUC Leu	CCC Pro	CAC His	CGC Arg	C
C	CUA Leu	CCA Pro	CAA Gln	CGA Arg	A
	CUG Leu	CCG Pro	CAG Gln	CGG Arg	G
	AUU Ile	ACU Thr	AAU Asn	AGU Ser	U
	AUC Ile	ACC Thr	AAC Asn	AGC Ser	C
A	AUA Ile	ACA Thr	AAA Lys	AGA Arg	A
	AUG Met	ACG Thr	AAG Lys	AGG Arg	G
	GUU Val	GCU Ala	GAU Asp	GGU Gly	U
	GUC Val	GCC Ala	GAC Asp	GGC Gly	C
G	GUA Val	GCA Ala	GAA Glu	GGA Gly	A
	GUG Val	GCG Ala	GAG Glu	GGG Gly	G

[a] All codons are read 5′ → 3′, i.e., 5′-pXpYpZ-3′.

stood in quiet repose, its pricey clothing shops beckoning the fashionable shopper. But here, in the "Colonnade," one could look down a tunnel the length of a football field and see in fluorescent-lit relief scores of open stalls offering everything from gourmet yogurt-covered pretzels to cotton candy, lasagna to knishes, and of course crabs. Upstairs the fast-food stalls emitted a greasy effluvium, but down here Rayna felt herself surrounded by the pleasant intermingling of exotic aromas. Why, she wondered, had she never come here before? The drive from the National Institutes of Health in Bethesda had taken just under an hour, and it had taken them almost that long again to find a parking space. But now she was ready to indulge herself in an array of gustatory delights.

The seminar at the NIH had gone well. A small but enthusiastic audience in the red-carpeted Masur Auditorium had listened to the first public presentation on "Biocryptography — New Age Secretion of Information", a title Rayna had invented with considerable pride, and not a little mischief. She had found it flattering that, during the question-and-answer period, several of her listeners had addressed her as Dr. Laszlo, and she had done nothing to alert them to the fact that the attainment of her degree remained some months in the future. The flush of success had intensified when, as she left the platform, an admirer furtively handed her a small envelope bearing the message "To Dr. Laszlo, with thanks" carefully inscribed with a fine calligraphy pen. Without reading the enclosed note, she had stuffed the envelope into her purse, met Sam, and smiled through the inevitable post-seminar chat with her hosts. They had seemed impressed and promised to invite her back again to expound further on her theories.

As they drove along the Capital Beltway and proceeded north on Interstate 95, the excitement of the seminar began to fade, and Rayna realized that she had consumed nothing all morning except for the coffee and doughnuts at the typical "social hour" that had preceded her talk. The rumbling in her stomach brought home the realization that she was famished.

Having rushed from one stall to the next, filling their trays with an eclectic sampling of foods, they searched in vain for a place to sit. Baltimore tradition apparently dictates eating while standing at one of the raised platforms scattered about the hall. Too hungry to care about amenities, Rayna and Sam found an empty platform and quickly began devouring their food.

Between the sushi and baklava, Rayna decided to return to a problem she and Sam had confronted a few days earlier.

"Sam, do you remember that jumble of data that came through the computer?"

"You mean the peptide experiment with the unusual data list? It was kind of strange. Some of the cleavage fragments were shown in upper case letters, some in lower case; some of the products were in Roman numerals, some in Arabic."

"I've been giving that some thought," Rayna said. "I think I've figured out what was going on. Look at this."

As Rayna pulled a few sheets of paper out of her purse Sam rolled his eyes, wondering if she were capable of leaving her job at the office. On the other hand, he did admire both her tenacity and her logical approach to problems.

"I think there are really two peptides here. I've resolved the data into two sets," she said, pointing to a dual list of fragments. "All we have to do is solve them."

"Great," said Sam, although not nearly as enthusiastically as Rayna would have hoped. "We'll just stand here munching our goodies and calculate away."

"C'mon Sam. Finish eating. There's a quiet place a couple of hundred yards from here. We'll sit overlooking the harbor. If you work on one peptide and I do the other, it shouldn't take too long."

"That's easy enough for you to say," he replied.

"I know you're not crazy about this idea," said Rayna. "But it will give you a good chance to practice the sequencing strategies we've been working on."

Making a desperate grab for the last of his food, Sam allowed Rayna to lead him by the hand, past the mob of diners. Outside, they worked their way through a crowd which had gathered to board and inspect a huge naval ship docked in the harbor, finally reaching a small flower garden, where they seated themselves on wooden benches and began to study the data.

"The funny thing about all this," Rayna began, "is that both peptides are exactly the same length. I don't know if that's significant or not. Why don't you try to solve peptide A, and I'll work on B."

Sam quickly scanned the data that Rayna gave him.

"Wait a minute," he said. "Most of this looks familiar, and I remember the rules you told me concerning the cleavage specificities. But what is this 'Cpdz/ $LiBH_4$' business?"

"It looks to me like a combination of carboxypeptidase and lithium borohydride. Apparently they remove the C-terminal amino acid with carboxypeptidase, then reduce the remaining peptide with lithium borohydride. After hydrolysis and amino acid analysis you get the amino acid just preceding the C-terminus.

"There's also a new wrinkle in the data for peptide B," Rayna continued. "Not only is there a carboxypeptidase/lithium borohydride step, but it looks like one of the peptic fragments is isolated, treated with amidase to convert an amide to the free acid, then redigested with pepsin."

"I guess that would convert either glutamine to glutamate or asparagine to aspartate," said Sam. "That would generate a new site for pepsin. I'll leave that to you. It looks like peptide A is about as much as I can handle."

The data for the two peptides are shown in Tables 6 and 7, respectively.

About an hour later Rayna looked over at Sam. Virtually immersed in a sea of yellow paper, he scratched away with his pencil, often grimacing at a particularly tricky section of the peptide. Most of the lunchtime strollers had returned to their offices, and the evening crowd had not yet arrived. A chill wind, foreshadowing weather to come, blew in off the harbor, and Rayna, sensing that Sam would need significantly more time, suggested that they leave.

"Let's go back to my place, Sam. You can finish this in the car, and I'll make some dinner for us."

The departure of Rayna's roommate for a week-long business trip allowed Sam to spread his papers around without fear of being called a slob and to work through the remainder of the sequence while Rayna prepared one of his favorite dishes. Over dinner they compared notes.

"Peptide A," Sam began. "Has 53 amino acid residues. I don't think I made any errors in the sequence, since everything seems consistent. I even translated it into the one letter code, but I can't make any sense out of it. I can make out specific words, things like LINKS, TREADS, RING, WANT, FEAR, PAY MY LASER, ICE, RINK, but how can they be arranged to yield a message?"

"That's the same problem I have," Rayna said. "Peptide B also has 53 amino acids. Maybe that indicates some connection between them. But look at what comes out. Things like REQWIRE, SAFER, CHLRIN, LINAGE, KNEW, GAL, MART, WARD, REIN. Even allowing for the probability that REQWIRE and CHLRIN represent 'require' and 'chlorine', I don't see any sense here."

Table 6

Sequence Data For Peptide A

Composition	(Ala$_7$,Arg$_5$,Asp$_6$,Cys,Glu$_6$,Gly,Ile$_4$,Leu$_3$,Lys$_3$,Met$_3$,Phe, Pro,Ser$_4$,Thr$_5$,Trp,Tyr$_2$)
LiBH$_4$	(2,6-diamino-1-hexanol)
CB1	(Ala,Arg,Asp$_2$,Gly,Hsr,Ile,Thr,Trp)
CB2	(Ala$_3$,Arg,Glu,Hsr,Phe,Pro,Thr,Tyr)
CB3	(Ala,Arg$_2$,Asp,Cys,Glu$_2$,Ile$_2$,Leu,Lys,Ser$_2$,Thr,Tyr)
CB4	(Ala$_2$,Arg,Asp$_3$,Glu$_3$,Hsr,Ile,Leu$_2$,Lys$_2$,Ser$_2$,Thr$_2$)
CT1	(Met,Tyr)
CT2	(Ala$_2$,Arg,Glu,Pro,Tyr)
CT3	(Ala$_2$,Asp,Met,Phe,Thr$_2$)
CT4	(Ala,Arg$_2$,Asp,Cys,Glu$_2$,Ile$_2$,Leu,Lys,Ser$_2$,Thr)
CT5	(Ala$_2$,Arg$_2$,Asp$_4$,Glu$_3$,Gly,Ile$_2$,Leu$_2$,Lys$_2$,Met,Ser$_2$,Thr$_2$,Trp)
E1	(Ala,Trp)
E2	(Ala,Arg,Pro)
E3	(Ala,Glu,Phe,Thr)
E4	(Ala,Asp,Met,Thr)
E5	(Ala,Leu,Met,Tyr$_2$)
E6	(Ala,Arg,Glu$_2$,Thr$_2$)
E7	(Ala,Asp$_2$,Ile,Leu,Lys,Ser)
E8	(Arg,Asp$_2$,Glu,Gly,Ile,Leu,Lys,Met,Ser)
E9	(Arg$_2$,Asp,Cys,Glu$_2$,Ile$_2$,Lys,Ser$_2$,Thr)
T1	(Asp,Ile,Lys)
T2	(Arg,Glu,Leu,Met)
T3	(Asp,Ile,Leu,Lys)
T4	(Ala,Asp,Glu,Lys,Ser)
T5	(Arg,Cys,Glu,Ile,Ser,Thr)
T6	(Ala,Arg,Asp,Glu,Ser,Thr$_2$)
T7	(Ala$_2$,Arg,Glu,Leu,Met,Pro,Ser,Tyr$_2$)
T8	(Ala$_3$,Arg,Asp$_2$,Glu,Gly,Ile,Met,Phe,Thr$_2$,Trp)
P1	Leu
P2	Phe
P3	Tyr
P4	(Ala,Glu)
P5	(Met,Tyr)
P6	(Ala,Leu,Ser)
P7	(Ala,Asp,Thr)
P8	(Asp,Lys,Ser)
P9	(Arg,Glu,Thr)
P10	(Arg,Cys,Glu,Ile)
P11	(Asp,Ile,Leu,Lys,Ser)
P12	(Ala$_2$,Arg,Glu,Pro)
P13	(Arg,Asp,Glu,Gly,Ile,Met)
P14	(Arg,Asp,Glu,Ile,Lys,Ser,Thr)
P15	(Ala$_2$,Asp,Met,Thr$_2$,Trp)

Table 6 (continued)

Sequence Data For Peptide A

Cpdz/$LiBH_4$	
CB1	2-amino-1,3-butanediol
E8	3-amino-4-hydroxybutanoate
T1	3-amino-4-hydroxybutanoate
T3	3-amino-4-hydroxybutanoate
T5	2-amino-1,3-butanediol
P10	2-amino-3-methyl-1-pentanol

Table 7

Sequence Data For Peptide B

Composition	(Ala_6,Arg_7,Asp_8,Cys,Glu_7,Gly_2,His,Ile_4,Leu_3,Lys,Met_3, Phe_2,Ser,Thr_2,Trp_3,Tyr_2)
CB1	(Ala,Phe,Thr)
CB2	(Ala,Asp,Glu,Gly,Leu,Lys,Hsr,Trp,Tyr)
CB3	(Ala_2,Arg_3,Asp_4,Glu,Ile,Hsr,Thr,Trp,Tyr)
CB4	(Ala_2,Arg_4,Asp_3,Cys,Glu_5,Gly,His,Hsr,Ile_3,Leu_2,Phe,Ser,Trp)
CT1	Tyr
CT2	Thr
CT3	(Ala,Arg,Asp,Tyr)
CT4	(Arg,Glu_2,Trp)
CT5	(Ala,Arg,Glu,Ile,Ser,Phe)
CT6	(Ala_2,Arg,Gly,Leu,Met,Thr,Trp)
CT7	(Ala,Arg,Asp_3,Glu,Ile,Met,Phe)
CT8	(Ala,Arg_2,Asp_4,Cys,Glu_3,Gly,His,Ile_2,Leu_2,Lys,Met,Trp)
T1	Arg
T2	(Arg,Asp_2,Tyr)
T3	(Ala,Arg,Thr,Trp)
T4	(Arg,Asp,Cys,His,Leu)
T5	(Arg,Glu_2,Ile,Trp)
T6	(Ala,Arg,Glu_2,Phe,Ser)
T7	(Ala,Asp_2,Glu,Ile,Met,Phe,Thr)
T8	(Ala_2,Arg,Asp,Glu,Gly,Leu,Met,Trp,Tyr)
T9	(Ala,Asp_2,Glu,Gly,Ile_2,Leu,Lys,Met)
E1	Ala
E2	Gly
E3	(Phe,Thr)
E4	(Ala,Leu,Met)
E5	(Ala,Arg,Thr,Trp)
E6	(Asp,Glu_2,Gly,Lys,Met,Trp,Tyr)
E7	(Ala,Arg_2,Glu_3,Ile,Ser,Trp)
E8	(Ala,Arg_2,Asp_4,Glu,Ile,Met,Tyr)
E9	(Ala,Arg_2,Asp_3,Cys,Glu,His,Ile_2,Leu_2,Phe)

Table 7 (continued)

Sequence Data For Peptide B

P1	Trp
P2	Phe
P3	Glu
P4	Asp
P5	Arg
P6	Glu_2
P7	(Phe,Thr)
P8	(Arg,Ile,Trp)
P9	(Ala,Arg,Trp)
P10	(Ala,Asp,Met)
P11	(Ala,Glu,Ser)
P12	(Ala,Gly,Tyr)
P13	(Arg,Asp,Tyr)
P14	(Asp,Glu,Ile)
P15	(Asp,Glu,Lys,Met)
P16	(Arg,Asp,Ile,Leu)
P17	(Ala,Arg,Leu,Met,Thr)
P18	(Arg,Asp,Cys,Glu,His)
P19	(Ala,Asp,Gly,Ile,Leu)
Amidase/Pepsin (P18)	
AP1	(Arg,Glu)
AP2	(Asp,Cys,His)
$LiBH_4$	
P14	3-amino-4-hydroxybutanoate
Cpdz/$LiBH_4$	
E9	3-amino-4-hydroxybutanoate
P16	2-amino-3-methyl-1-pentanol
AP2	2-amino-3-mercapto-1-propanol

Sam looked up with a puzzled frown.

"What's that music?"

"Oh, the soundproofing in this place is just awful," she replied. "That must be Michael, next door. He's a musician. Why don't I invite him in for dessert? Maybe someone with an artistic personality can help make sense of all this."

Michael greeted Sam amiably and, as they all shared coffee, looked with dismay at the papers spread over the table.

"This doesn't mean a whole lot to me," he said. "But if I were presented with a bunch of musical fragments I'd try first to combine them, then to mix them up, sort of taking a piece from here, a piece from here, until something made sense."

"Combining the sequences might help," said Sam enthusiastically. "Look, each of the peptides has a single cysteine residue. Suppose this is really a single protein with two chains joined by a disulfide bond. Maybe we should line up the sequences with the cysteines opposite each other and see if anything falls out."

Rayna dug into one of the kitchen cabinet drawers, extracting a small pair of scissors. Sam found some tape and began cutting and pasting the two sequences. Finally he looked up, dejected.

"I don't think that was such a hot idea. I tried reading across the disulfide bond in all possible directions, but I still can't read anything sensible. What's that?" he asked Rayna, as she began opening a small envelope.

"I'm not really sure. A strange little man handed it to me after the seminar, and then he immediately disappeared. He didn't say anything, and I had the distinct impression he was trying to avoid letting me see his face."

She removed a small index card from the envelope.

"This is weird," she said, dropping the card on the table for the two men to see.

"'Greetings from Exxon'!" Michael exclaimed incredulously. "What kind of people come to your seminars, anyhow? And look at this. He gave you his phone numbers. 313–164–2213, and 242–122–5122(4). I guess the first one is his home number and the other is at work. The (4) must be an extension number. Do you know anything about this guy?"

"Not a thing," Rayna replied. "I can't believe someone would just expect me to call him. There's no name. Nothing."

Rayna's voice began to fade as her face assumed a look of intense concentration. Michael, never having seen her like this, shot a quizzical glance at Sam. The latter, more familiar with Rayna's work habits, nodded as if to suggest that the two men leave her.

Having nothing better to do, Sam and Michael began cleaning up the dinner dishes while Rayna, oblivious to their activity, wrote rapidly and began to smile knowingly. Finally she looked up, consulted the calendar, and sighed.

"There's something peculiar about these numbers. If you add up all the digits you come up with 53, exactly the number of amino acid residues in each peptide. But they're not phone numbers. The clue is Exxon. Michael, you would have no way to know this, but DNA often has regions called introns and exons. Introns, or intervening regions, seem to be a sort of filler, while exons, or expressed regions, contain meaningful genetic information. That's what he's telling us here. The numbers represent regions of the sequences that should be read to get a meaningful message. Apparently you just take an exon from one strand, follow it with an exon from the other, and continue jumping back and forth, until the whole message pops out. Fortunately, it's not even Christmas yet, so we have plenty of time to report this to the Director.

"Here," she continued. "Let me show you the whole thing."

"Wait a minute," Michael interjected. "Before you do that, how about showing me how you got this jumble of letters in the first place."

Rayna smiled. "I'll let Sam tell you about his peptide first. Go on Sam."

Solution

Rayna brought a stack of fan-folded computer paper to the table.

"I generally save my printer output, Sam. The back side is usually clean enough for scratch work. You can use it to show us how you solved the sequence."

"Jumping right in," Sam said, "let's begin at the end. The first thing we see is the product of lithium borohydride reduction followed by hydrolysis. From this we deduce that the *C*-terminus is lysine.

"Of the four cyanogen bromide fragments, CB3 has no methionine and must therefore be at the *C*-terminus. Consequently, its *C*-terminal amino acid is lysine, and we can show the partial sequence as in Figure 10.

"Notice the sequence at the *C*-terminus of CB1. We can place methionine from the specificity of cyanogen bromide, and threonine from the carboxypeptidase/lithium borohydride experiment. But that's as far as we can go with the cyanogen bromide data.

"Now let's look at the chymotrypsin fragments. CT4, of course, must be at the *C*-terminus, so a partial sequence is as shown in Figure 11.

"CB3 contains residues 39 to 53, while CT4 contains residues 40 to 53. The only amino acid present in CB3 that's not in CT4 is tyrosine, which must then be at the *N*-terminus of CB3 (position 39):

CB3 = Tyr–(Ala,Arg_2,Asp,Cys,Glu_2,Ile_2,Leu,Ser_2,Thr)–Lys

"Since CB3 is a cyanogen bromide fragment, the tyrosine at its *N*-terminus must be preceded by methionine; the sequence Met–Tyr is in CT1, which therefore must precede CT4:

CT1–CT4 (residues 38 to 53) =
Met–Tyr–(Ala,Arg_2,Asp,Cys,Glu_2,Ile_2,Leu,Ser_2,Thr)–Lys

"We can extend the partial chymotryptic sequence to show Figure 12.

"We can determine the *N*-terminal cyanogen bromide fragment, but this part is a little tricky. In CT3 methionine precedes phenylalanine. If CB2 were the *N*-terminal fragment, then methionine would not precede phenylalanine; thus CB2 cannot be the *N*-terminal.

"Similarly, in CT5, methionine precedes tryptophan. If CB1 were the *N*-terminal fragment, then methionine would not precede tryptophan; thus CB1 cannot be the *N*-terminal.

"CB4 is the only remaining cyanogen bromide fragment; it must be the *N*-terminal, so that we can show the cyanogen bromide sequence as in Figure 13.

"Fragment CB4 contains amino acid residues 1 to 19. The first methionine in the peptide, then, is at position 19. Only two remaining chymotryptic fragments contain methionine. If CT3 were at the *N*-terminus, then methionine would be one of the first six amino acids. Since we know that the first methionine doesn't appear until position 19, CT3 cannot be at the *N*-terminus. Thus CT5 must be *N*-terminal, and the chymotryptic sequence is as shown in Figure 14.

"CT5 contains the first 24 residues. The first 19 of them are also found in CB4. If we subtract these from CT5 we get Figure 15.

"The peptide contains only one residue each of glycine and tryptophan; these are both in the pentapeptide following $methionine_{19}$. Both of them are also in cyanogen bromide fragment CB1, therefore, it must follow CB4; the sequence is

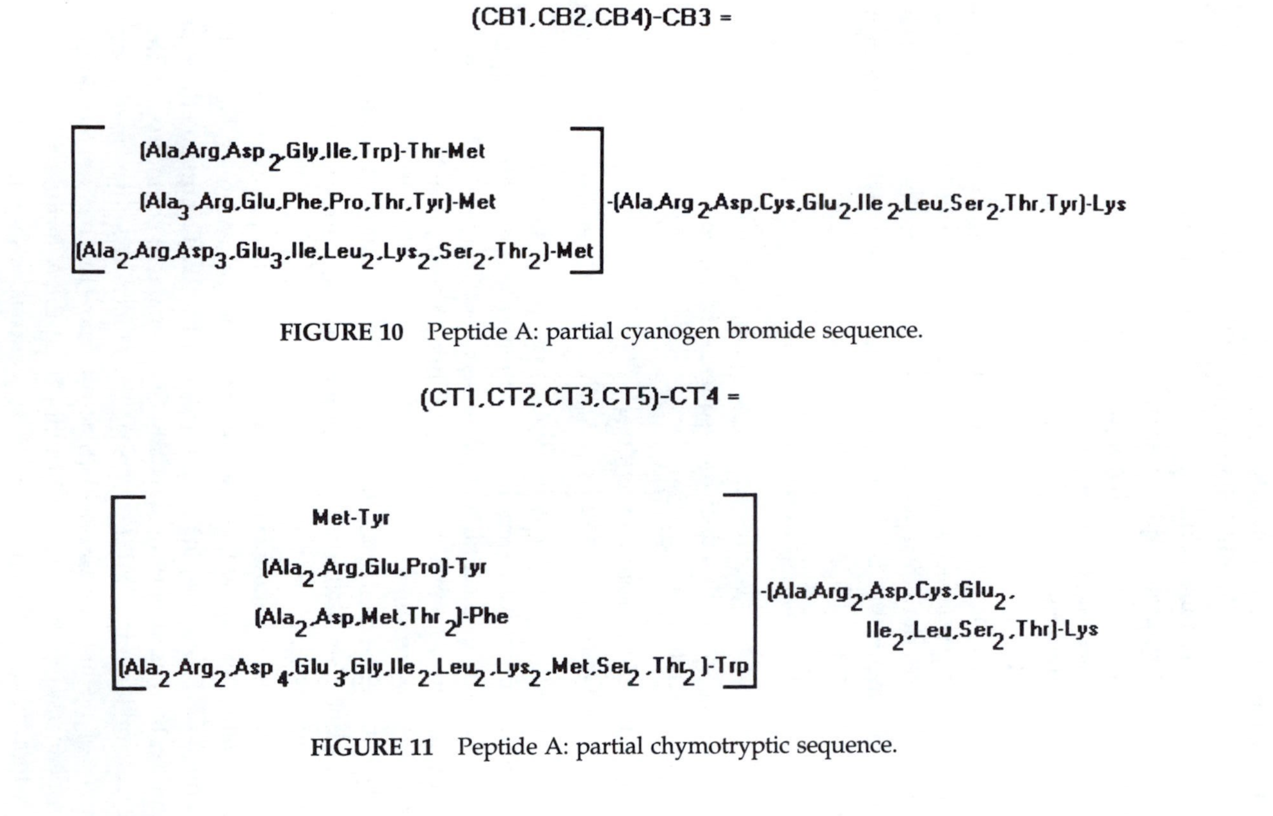

FIGURE 10 Peptide A: partial cyanogen bromide sequence.

FIGURE 11 Peptide A: partial chymotryptic sequence.

(CT2,CT3,CT5)-CT1-CT4 =

[(Ala$_2$,Arg,Glu,Pro)-Tyr
(Ala$_2$,Asp,Met,Thr$_2$)-Phe
(Ala$_2$,Arg$_2$,Asp$_4$,Glu$_3$,Gly,Ile$_2$,Leu$_2$,Lys$_2$,Met,Ser$_2$,Thr$_2$)-Trp]-Met-Tyr-(Ala,Arg$_2$,Asp,Cys,Glu$_2$,
Ile$_2$,Leu,Ser$_2$,Thr)-Lys

FIGURE 12 Peptide A: partial chymotryptic sequence.

CB4-(CB1,CB2)-CB3 =

(Ala$_2$,Arg,Asp$_3$,Glu$_3$,
Ile,Leu$_2$,Lys$_2$,Ser$_2$,Thr$_2$)-Met-[(Ala,Arg,Asp$_2$,Gly,Ile,Trp)-Thr-Met
(Ala$_3$,Arg,Glu,Phe,Pro,Thr,Tyr)-Met]-Tyr-(Ala,Arg$_2$,Asp,Cys,Glu$_2$,
Ile$_2$,Leu,Ser$_2$,Thr)-Lys

FIGURE 13 Peptide A: partial cyanogen bromide sequence.

CT5-(CT2,CT3)-CT1-CT4 =

(Ala$_2$,Arg$_2$,Asp$_4$,Glu$_3$,
Gly,Ile$_2$,Leu$_2$,Lys$_2$,Met,Ser$_2$,Thr$_2$)-Trp-[(Ala$_2$,Arg,Glu,Pro)-Tyr / (Ala$_2$,Asp,Met,Thr$_2$)-Phe]-Met-Tyr-(Ala,Arg$_2$,Asp,Cys,Glu$_2$,
Ile$_2$,Leu,Ser$_2$,Thr)-Lys

FIGURE 14 Peptide A: partial chymotryptic sequence.

CT5-(CT2,CT3)-CT1-CT4 =

(Ala$_2$,Arg,Asp$_3$,Glu$_3$,Ile,Leu$_2$,Lys$_2$,
Ser$_2$,Thr$_2$)-Met-(Arg,Asp,Gly,Ile)-Trp-[(Ala$_2$,Arg,Glu,Pro)-Tyr / (Ala$_2$,Asp,Met,Thr$_2$)-Phe]-Met-Tyr-(Ala,Arg$_2$,Asp,Cys,Glu$_2$,
Ile$_2$,Leu,Ser$_2$,Thr)-Lys

FIGURE 15 Peptide A: partial chymotryptic sequence.

CB4–CB1–CB2–CB3 =
(Ala$_2$,Arg,Asp$_3$,Glu$_3$,Ile,Leu$_2$,Lys$_2$,Ser$_2$,Thr$_2$)–Met–(Arg,Asp,Gly,Ile)–Trp–
(Ala,Asp)–Thr–Met–(Ala$_3$,Arg,Glu,Phe,Pro,Thr,Tyr)–Met–Tyr–
(Ala,Arg$_2$,Asp,Cys,Glu$_2$,Ile$_2$,Leu,Ser$_2$,Thr)–Lys

"Tryptophan$_{24}$ is followed by (Ala,Asp)–Thr–Met; since this grouping is found only in CT3, CT3 must follow CT5 and the entire sequence of chymotryptic fragments is

CT5–CT3–CT2–CT1–CT4 =
(Ala$_2$,Arg,Asp$_3$,Glu$_3$,Ile,Leu$_2$,Lys$_2$,Ser$_2$,Thr$_2$)–Met–(Arg,Asp,Gly,Ile)–Trp–
(Ala,Asp)–Thr–Met–(Ala,Thr)–Phe–(Ala$_2$,Arg,Glu,Pro)–Tyr–Met–Tyr–
(Ala,Arg$_2$,Asp,Cys,Glu$_2$,Ile$_2$,Leu,Ser$_2$,Thr)–Lys

"Now let's consider the elastase fragments. The only tryptophan in the original peptide is in fragment E1 = Trp–Ala (residues 24 to 25). Residues 26 to 28 = Asp–Thr–Met must be part of a tetra- or pentapeptide; only E4 qualifies: E4 = Asp–Thr–Met–Ala. The sequence Thr–Phe (residues 30 to 31) is from E3: E3 = Thr–Phe–Glu–Ala (residues 30 to 33). The proline following phenylalanine$_{31}$ is in E2 = (Arg,Pro)–Ala (residues 34 to 36), and the sequence Tyr–Met–Tyr (residues 37 to 39) is found in E5: E5 = Tyr–Met–Tyr–Leu–Ala (residues 37 to 41). We can show a partial sequence for residues 24 to 41:

E1–E4–E3–E2–E5 =
Trp–Ala–Asp–Thr–Met–Ala–Thr–Phe–Glu–Ala–(Arg,Pro)–Ala–
Tyr–Met–Tyr–Leu–Ala

The only remaining methionine (residue 19) is in E8 = (Asp,Glu,Leu,Lys,Ser)–Met–(Arg,Ile)–Asn–Gly (residues 14 to 23). We know asparagine at position 22 from the carboxypeptidase/lithium borohydride data.

"E9 is the *C*-terminal fragment (residues 42 to 53): E9 = (Arg$_2$,Asp,Cys,Glu$_2$, Ile$_2$,Ser$_2$,Thr)–Lys. The only remaining elastase fragments are E6 and E7, which must be at the *N*-terminus; the entire sequence is shown in Figure 16.

"Now we can look at the tryptic fragments. Positions 34 and 35 contain proline and arginine. If proline preceded arginine, then proline and phenylalanine would be in the same tryptic fragment. Since they are not, we conclude that arginine precedes proline, and proline is the *N*-terminus of T7: T7 = Pro–Ala–Tyr–Met–Tyr–Leu–Ala–(Glu,Ser)–Arg (residues 35 to 44).

"Since T8 contains phenylalanine, it must come just before T7: T8 = Ile–Asn–Gly–Trp–Ala–Asp–Thr–Met–Ala–Thr–Phe–Glu–Ala–Arg (residues 21 to 34). The remaining methionine (residue 19) is in T2: T2 = (Glu,Leu)–Met–Arg (residues 17 to 20). The partial tryptic sequence so far (residues 17 to 44) is

T2–T8–T7 =
(Glu,Leu)–Met–Arg–Ile–Asn–Gly–Trp–Ala–Asp–Thr–Met–Ala–Thr–Phe–
Glu–Ala–Arg–Pro–Ala–Tyr–Met–Tyr–Leu–Ala–(Glu,Ser)–Arg

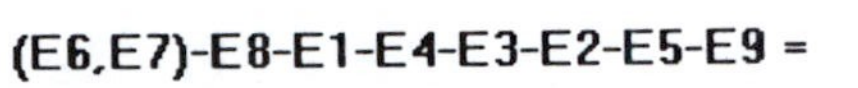

$$\left[\begin{array}{c}(Arg,Glu_2,Thr_2)\text{-}Ala \\ (Asp_2,Ile,Leu,Lys,Ser)\text{-}Ala\end{array}\right]$$-(Asp,Glu,Leu,Lys,Ser)-Met-(Arg,Ile)-Asn-Gly-Trp-Ala-Asp-Thr-Met-Ala-Thr-

Phe-Glu-Ala-(Arg,Pro)-Ala-Tyr-Met-Tyr-Leu-Ala-(Arg_2,

Asp,Cys,Glu_2,Ile_2,Ser_2,Thr)-Lys

FIGURE 16 Peptide A: partial elastase sequence

"Since E8 contained residues 14 to 23, we can now show it as (Asp,Lys,Ser)–(Glu,Leu)–Met–Arg–Ile–Asn–Gly. The lysine in E8 must be the *C*-terminus of the tryptic fragment preceding T2. Thus positions 14 to 16, (Asp,Ser)–Lys, must come from T4 = (Ala,Glu)–(Asp,Ser)–Lys (residues 12 to 16). Residues 12 to 13, (Ala,Glu), must come from fragment E6, since E7 contains no glutamate: E6 = (Arg,Glu,Thr_2)–Glu–Ala (residues 8 to 13). As a result, E7 must be the *N*-terminal elastase fragment: E7 = (Asp_2,Ile,Leu,Lys,Ser)–Ala (residues 1 to 7).

"The first 20 residues are (Asp_2,Ile,Leu,Lys,Ser)–Ala–(Arg,Glu,Thr_2)–Glu–Ala–(Asp,Ser)–Lys–(Glu,Leu)–Met–Arg. Residues 12 to 16, Glu–Ala–(Asp,Ser)–Lys, are found in T4; residues 7 to 11, Ala–(Arg,Glu,Thr_2), must be from T6, since T5 has one glutamate but no alanine. Thus, T6 = (Asp,Ser)–Ala–(Glu,Thr_2)–Arg (residues 5 to 11).

"Residues 1 to 4 must correspond to T3: T3 = (Ile,Leu)–Asn–Lys. The remaining lysine is in T1, which is the *C*-terminal fragment: T1 = Ile–Asn–Lys (residues 51 to 53). Fragment T5 must therefore reside between T7 and T1: T5 = (Cys,Glu,Ile,Ser)–Thr–Arg (residues 45 to 50). The amino acids just prior to the *C*-termini of T1, T3, and T5 are all known from the carboxypeptidase/lithium borohydride data.

"The entire tryptic sequence is

T3–T6–T4–T2–T8–T7–T5–T1 =
(Ile,Leu)–Asn–Lys–(Asp,Ser)–Ala–(Glu,Thr_2)–Arg–Glu–Ala–(Asp,Ser)–Lys–(Glu,Leu)–Met–Arg–Ile–Asn–Gly–Trp–Ala–Asp–Thr–Met–Ala–Thr–Phe–Glu–Ala–Arg–Pro–Ala–Tyr–Met–Tyr–Leu–Ala–(Glu,Ser)–Arg–(Cys,Glu,Ile,Ser)–Thr–Arg–Ile–Asn–Lys

"We're almost done. Look at the peptic fragments. If residue 1 were isoleucine, it would appear as a free amino acid in the peptic digest; since it does not, the first two positions must be Leu–Ile. Consequently we can show positions 1 to 6 as Leu–Ile–Asn–Lys–(Asp,Ser). The leucine, isoleucine, lysine, and serine in this grouping should appear in either a tetrapeptide (if serine is residue 6) or a pentapeptide (if serine is residue 5). Fragment P11 satisfies these requirements: P11 = Leu–Ile–Asn–Lys–Ser (residues 1 to 5).

"Residues 6 to 10 are Asp–Ala–(Glu,Thr_2). Asp–Ala will appear in a dipeptide if glutamate is residue 8, a tripeptide if glutamate is residue 9, or a tetrapeptide if glutamate is residue 10. We find P7 = Asp–Ala–Thr (residues 6 to 8).

"Residues 9 to 11 (Glu,Thr)–Arg are found in P9. The first 11 residues are

P11–P7–P9 = Leu–Ile–Asn–Lys–Ser–Asp–Ala–Thr–Glu–Thr–Arg

"Since there is no tripeptide containing glutamate, alanine, and serine, residues 12 to 13 must come from the dipeptide P4: P4 = Glu–Ala. Similarly, since there is no tetrapeptide containing aspartate, serine, lysine, and leucine, residues 14 to 16 must come from the tripeptide P8: P8 = Asp–Ser–Lys.

"If residue 17 were glutamate we would find free glutamate among the peptic fragments; since we do not, leucine (P1) must be residue 17.

"The sequence of residues 18 to 23 is now known, and must come from P13: P13 = Glu–Met–Arg–Ile–Asn–Gly. Residues 24 to 25, Trp–Ala, are contained in P15: P15 = Trp–Ala–Asn–Thr–Met–Ala–Thr.

"Residue 31 is Phe = P2, and residues 32 to 36, Glu–Ala–Arg–Pro–Ala, are in P12. The sequence Tyr–Met (residues 37 to 38) is found in P5, and the remaining tyrosine at residue 39 is in P3. Residues 40 to 41, Leu–Ala, can come only from P6: P6 = Leu–Ala–Ser (residues 40 to 42).

"We can now show the first 42 residues:

P11–P7–P9–P4–P8–P1–P13–P15–P2–P12–P5–P3–P6 =
Leu–Ile–Asn–Lys–Ser–Asp–Ala–Thr–Glu–Thr–Arg–Glu–Ala–Asp–Ser–Lys–
Leu–Glu–Met–Arg–Ile–Asn–Gly–Trp–Ala–Asn–Thr–Met–Ala–Thr–Phe–Glu–
Ala–Arg–Pro–Ala–Tyr–Met–Tyr–Leu–Ala–Ser

"Of the remaining fragments (P10 and P14) only P14 contains lysine, which we know is the *C*-terminal residue. That means that P10 must precede P14. Once again we know the amino acid just prior to the *C*-terminus of P10 from the carboxypeptidase/lithium borohydride data: P10 = Glu–Arg–Ile–Cys.

"The remaining residues, Glu–Ser–Thr–Arg–Ile–Asn–Lys, are found in P14, and the entire sequence is

P11–P7–P9–P4–P8–P1–P13–P15–P2–P12–P5–P3–P6–P10–P14 =
Leu–Ile–Asn–Lys–Ser–Asp–Ala–Thr–Glu–Thr–Arg–Glu–Ala–Asp–Ser–
Lys–Leu–Glu–Met–Arg–Ile–Asn–Gly–Trp–Ala–Asn–Thr–Met–Ala–Thr–
Phe–Glu–Ala–Arg–Pro–Ala–Tyr–Met–Tyr–Leu–Ala–Ser–Glu–Arg–Ile–
Cys–Glu–Ser–Thr–Arg–Ile–Asn–Lys

If we convert to the single letter code we get LINKSDATETREADSKLE-MRINGWANTMATFEARPAYMYLASERICESTRINK." Sam turned to Rayna with a grin on his face. "How was that?" he asked.

"Super, Sam. I'm proud of you. While you were explaining the sequence to Michael I drew this cleavage map to summarize everything."

Rayna placed the diagram (Figure 17) on the table as she prepared to explain peptide B.

"In peptide B there's no end group analysis, as there was in peptide A, so we don't know either the *N*-terminal or the *C*-terminal amino acid. But we can look at the cyanogen bromide data and tell immediately that CB1 is the *C*-terminal fragment, since it lacks homoserine. There's nothing else we can say yet, so let's show a partial sequence (Figure 18).

"When we examine the chymotryptic fragments we see that CT2 (Thr) is the *C*-terminus. That means that the *C*-terminal cyanogen bromide fragment, CB1, is (Ala,Phe)–Thr; this must be preceded by methionine. Of the two phenylalanine-containing chymotryptic fragments, only CT7 contains methionine. Thus, CT7 must directly precede CT2:

CT7–CT2 = (Arg,Asp_3,Glu,Ile)–Met–Ala–Phe–Thr (residues 44 to 53)

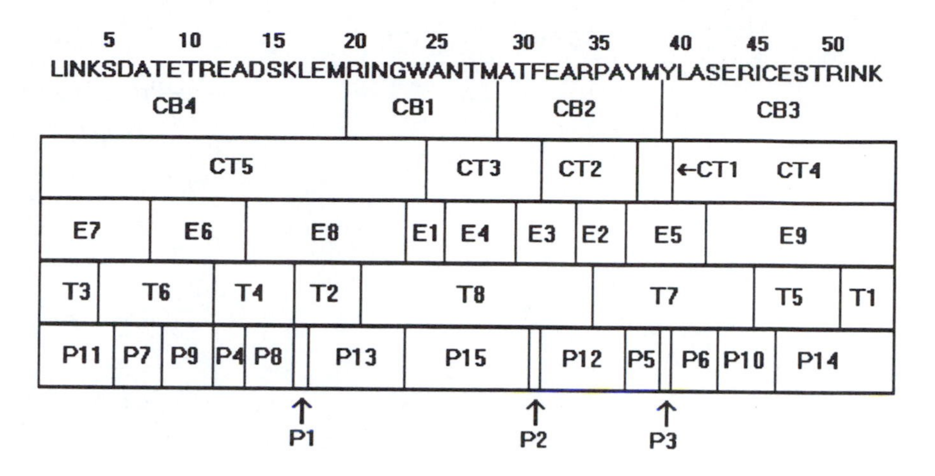

FIGURE 17 Cleavage map for peptide A.

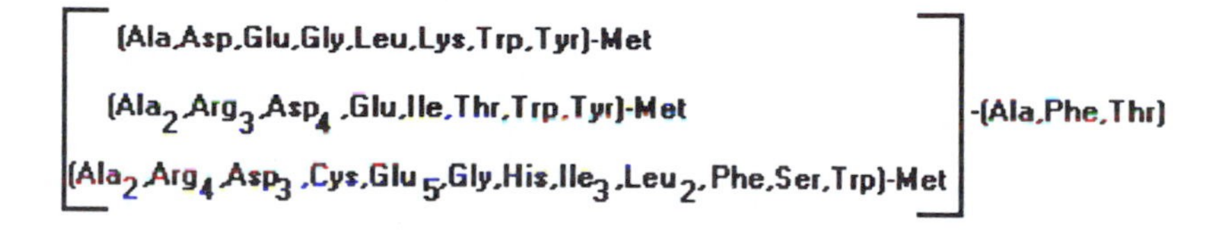

FIGURE 18 Peptide B: partial cyanogen bromide sequence.

The sequence (Arg,Asp_3,Glu,Ile) comes from either CB3 or CB4, so that one of these must directly precede CB1. "The partial chymotryptic sequence is shown in Figure 19.

"Now let's look at the tryptic fragments. T7 has no basics, so it must be the *C*-terminal fragment: T7 = (Asp_2,Glu,Ile)–Met–Ala–Phe–Thr (residues 46 to 53). That means that the chymotryptic sequence CT7–CT2 is (Asp,Arg)–(Asp_2,Glu,Ile)–Met–Ala–Phe–Thr. But, since position 45 must be the *C*-terminus of a tryptic fragment:

CT7–CT2 =
Asp–Arg–(Asp_2,Glu,Ile)–Met–Ala–Phe–Thr (residues 44 to 53)

The partial tryptic sequence is shown in Figure 20."

Michael interrupted. "It doesn't look like you've been able to draw too many conclusions about this peptide yet."

"Wait," Rayna replied. "And watch. Once we start looking at the elastase fragments, things will fall into place fast enough to make your head spin."

"It's doing that already," he said.

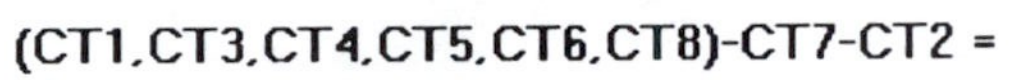

(CT1,CT3,CT4,CT5,CT6,CT8)-CT7-CT2 =

$$\left[\begin{array}{c} \text{Tyr} \\ \text{(Ala,Arg,Asp)-Tyr} \\ \text{(Arg,Glu}_2\text{)-Trp} \\ \text{(Ala,Arg,Glu,Ile,Ser)-Phe} \\ \text{(Ala}_2\text{,Arg,Gly,Leu,Met,Thr)-Trp} \\ \text{(Ala,Arg}_2\text{,Asp}_4\text{,Cys,Glu}_3\text{,Gly,His,Ile}_2\text{,Leu}_2\text{,Lys,Met)-Trp} \end{array}\right]\text{-(Arg,Asp}_3\text{,Glu,Ile)-Met-Ala-Phe-Thr}$$

FIGURE 19 Peptide B: partial chymotryptic sequence.

(T1,T2,T3,T4,T5,T6,T8,T9)–T7 =

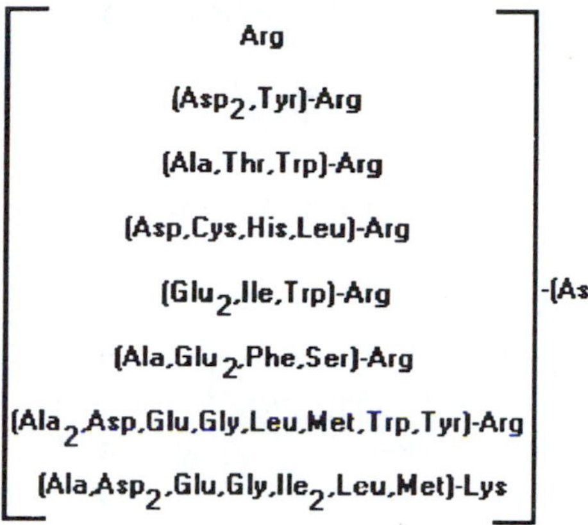

-(Asp2,Glu,Ile)-Met-Ala-Phe-Thr

FIGURE 20 Peptide B: partial tryptic sequence.

"Elastase fragment E3 (Phe–Thr) is the *C*-terminus. The remainder of the *C*-terminal sequence, residues 46 to 51, (Asp_2,Glu,Ile)–Met–Ala, comes from E8 = (Arg_2,Asp_2,Tyr)–(Asp_2,Glu,Ile)–Met–Ala."

"So E8 actually contains residues 41 to 51," interjected Michael. "What scares me is that I think I'm beginning to follow some of this."

Rayna continued. "We've already established cyanogen bromide fragment CB1 at the *C*-terminus, and we've seen that either CB3 or CB4 must directly precede CB1. Of these, only CB3 contains tyrosine, thus CB3 (residues 36 to 50) = (Ala_2,Arg,Thr,Trp)–(Arg,Asp,Tyr)–Asp–Arg–(Asp_2,Glu,Ile)–Met. The tetrapeptide (Asp,Tyr)–Asp–Arg (residues 42 to 45) must come from T2. That means that CB3 is (Ala_2,Arg,Thr,Trp)–Arg–(Asp,Tyr)–Asp–Arg–(Asp_2,Glu,Ile)–Met.

"The hexapeptide at the *N*-terminus of CB3 (residues 36 to 41) contains the four amino acids found in T3, so we can show the three tryptic fragments at the *C*-terminus:

T3–T2–T7 =
(Ala,Thr,Trp)–Arg–(Asp,Tyr)–Asp–Arg–(Asp_2,Glu,Ile)–Met–Ala–Phe–Thr
(residues 38 to 53)

The tyrosine preceding $aspartate_{44}$ comes from CT3 = Ala–Arg–Asp–Tyr. The grouping (Thr,Trp) at positions 38 to 39 comes from CT6 = (Ala,Gly,Leu,Met)–Ala–Arg–Thr–Trp."

"Hold it," Sam said. "I can see how you got Thr–Trp at the *C*–terminus of CT6, but how can you place the sequence Ala–Arg just before threonine?"

Rayna looked up from her notes.

"Fragment T3 contains residues 38 to 41. That means that residue 37 must be the *C*-terminus of another tryptic fragment, so it had to be a basic residue, namely arginine."

Sam and Michael stared at each other, the latter letting out a slow whistle. Rayna went on.

"The four chymotryptic fragments at the *C*-terminus (residues 32 to 53) are

CT6–CT3–CT7–CT2 =
(Ala,Gly,Leu,Met)–Ala–Arg–Thr–Trp–Ala–Arg–Asp–Tyr–Asp–Arg–
(Asp_2,Glu,Ile)–Met–Ala–Phe–Thr

Residues 32 to 37 come from T8, which actually contains residues 28 to 37: T8 = (Asp,Glu,Trp,Tyr)–(Ala,Gly,Leu,Met)–Ala–Arg, so that the *C*-terminal tryptic peptide is T8–T3–T2–T7 (residues 28 to 53): (Asp,Glu,Trp,Tyr)–(Ala,Gly,Leu,Met)–Ala–Arg–Thr–Trp–Ala–Arg–Asp–Tyr–Asp–Arg–(Asp_2,Glu,Ile)–Met–Ala–Phe–Thr.

"In residues 28 to 31 (Asp,Glu,Trp,Tyr), either tryptophan or tyrosine must be at the *C*-terminus, since residue 32 is the *N*-terminus of CT6. The only possibility is (Asp,Glu)–Trp followed by the single tyrosine of CT1 at position 31. "The three amino acids at positions 28 to 30, (Asp,Glu)–Trp, come from CT8, which contains residues 11 to 30. CT8 = (Ala,Arg_2,Asp_3,Cys,Glu_2,Gly,His,Ile_2,Leu_2,Lys,Met)–(Asp,Glu)–Trp.

"We can now show residues 11 to 53 in the partial chymotryptic sequence:

CT8–CT1–CT6–CT3–CT7–CT2 =
(Ala,Arg_2,Asp_3,Cys,Glu_2,Gly,His,Ile_2,Leu_2,Lys,Met)–(Asp,Glu)–Trp–Tyr–
(Ala,Gly,Leu,Met)–Ala–Arg–Thr–Trp–Ala–Arg–Asp–Tyr–Asp–Arg–
(Asp_2,Glu,Ile)–Met–Ala–Phe–Thr

"If we recognize that $alanine_{36}$ is the *N*-terminus of CB3, it becomes clear that residue 35 must be methionine.

"The grouping Trp–Tyr (residues 30 to 31) must come from CB2, since CB4 contains no tyrosine. So CB2 = Lys–(Asp,Glu)–Trp–Tyr–(Ala,Gly,Leu)–Met. Residues 11 to 53 in the partial chymotryptic sequence can now be shown as

CT8–CT1–CT6–CT3–CT7–CT2 =
(Ala,Arg_2,Asp_3,Cys,Glu_2,Gly,His,Ile_2,Leu_2,Met)–Lys–(Asp,Glu)–Trp–Tyr–
(Ala,Gly,Leu)–Met–Ala–Arg–Thr–Trp–Ala–Arg–Asp–Tyr–Asp–Arg–
(Asp_2,Glu,Ile)–Met–Ala–Phe–Thr

"In the cyanogen bromide sequence, CB4 is the *N*-terminal fragment, so we can show

CB4–CB2–CB3–CB1 =
(Ala,Arg_2,Glu_3,Ile,Phe,Ser,Trp)–(Ala,Arg_2,Asp_3,Cys,Glu_2,Gly,His,Ile_2,Leu_2)–
Met–Lys–(Asp,Glu)–Trp–Tyr–(Ala,Gly,Leu)–Met–Ala–Arg–Thr–Trp–Ala–
Arg–Asp–Tyr–Asp–Arg–(Asp_2,Glu,Ile)–Met–Ala–Phe–Thr

"Positions 1 to 10 contain CT4 = (Arg,Glu_2)–Trp and CT5 = (Ala,Arg,Glu, Ile,Ser)–Phe. The chymotryptic sequence can be shown as in Figure 21.

(CT4,CT5)-CT8-CT1-CT6-CT3-CT7-CT2 =

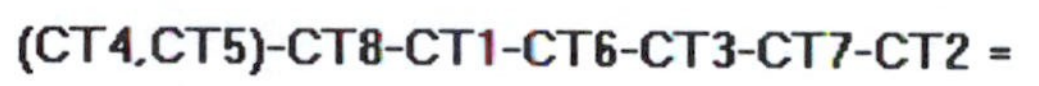

[(Arg,Glu_2)-Trp / (Ala,Arg,Glu,Ile,Ser)-Phe] -(Ala,Arg_2,Asp_3,Cys,Glu_2,Gly,His,Ile_2,Leu_2)-Met-Lys-(Asp,Glu)-Trp-

Tyr-(Ala,Gly,Leu)-Met-Ala-Arg-Thr-Trp-Ala-Arg-Asp-Tyr-

Asp-Arg-(Asp_2,Glu,Ile)-Met-Ala-Phe-Thr

FIGURE 21 Peptide B: partial chymotryptic sequence.

"The only lysine in the peptide, lysine$_{27}$, comes from tryptic fragment T9 = (Ala,Asp$_2$,Glu,Gly,Ile$_2$,Leu)–Met–Lys. Thus residues 11 to 27 = (Arg$_2$,Asp, Cys,Glu,His,Leu)–(Ala,Asp$_2$,Glu,Gly,Ile$_2$,Leu)–Met–Lys.

"Since residue 18 is the *N*-terminus of T9, residue 17 must be arginine, so we can rewrite residues 11 to 27 as (Arg,Asp,Cys,Glu,His,Leu)–Arg–(Ala,Asp$_2$,Glu,Gly,Ile$_2$,Leu)–Met–Lys.

"Now, watch how we find the *N*-terminal tryptic fragment. The phenylalanine in CT5 must be either at position 6 or at position 10; this phenylalanine is contained in tryptic fragment T6. If T6 were *N*-terminal then phenylalanine would be no further from the *N*-terminus than position 5. Consequently, T6 cannot be the *N*-terminal tryptic fragment.

"The tryptophan in CT4 must be either at position 4 or at position 10. Regardless of which chymotryptic fragment (CT4 or CT5) is at the *N*-terminus, at least one arginine residue must precede tryptophan. This tryptophan is contained in tryptic fragment T5. Since there must be an arginine preceding tryptophan, T5 cannot be the *N*-terminal tryptic fragment.

"T4 contains histidine, which is not found in either CT4 or CT5 (the two possible *N*-terminal chymotryptic fragments). Consequently, T4 cannot be the *N*-terminal tryptic fragment. The only remaining tryptic fragment is T1 = Arg, which must therefore be the *N*-terminus. A partial tryptic sequence (residues 1 to 27) is shown in Figure 22.

T1-(T4,T5,T6)-T9 =

$$\text{Arg-}\begin{bmatrix}\text{(Asp,Cys,His,Leu)-Arg}\\ \text{(Glu}_2\text{,Ile,Trp)-Arg}\\ \text{(Ala,Glu}_2\text{,Phe,Ser)-Arg}\end{bmatrix}\text{-(Ala,Asp}_2\text{,Glu,Gly,Ile}_2\text{,Leu)-Met-Lys}$$

FIGURE 22 Peptide B: partial tryptic sequence.

"From the chymotryptic sequence we saw earlier, it is clear that the tryptophan- and phenylalanine-containing tryptic fragments (T5 and T6, respectively) must precede T4; thus the *N*-terminal tryptic sequence is as shown in Figure 23.

T1-(T5,T6)-T4 =

$$\text{Arg-}\begin{bmatrix}\text{(Glu}_2\text{,Ile,Trp)-Arg}\\ \text{(Ala,Glu}_2\text{,Phe,Ser)-Arg}\end{bmatrix}\text{-(Asp,Cys,His,Leu)-Arg}$$

FIGURE 23 Peptide B: *N*-terminal tryptic sequence.

"Elastase fragments E8 and E3 have been established at the *C*-terminus:

E8–E3 = Arg–Asp–Tyr–Asp–Arg–(Asp$_2$,Glu,Ile)–Met–Ala–Phe–Thr (residues 41 to 53)

"Working back toward the *N*-terminus, the next four elastase fragments are

E5 = Arg–Thr–Trp–Ala (residues 37 to 40)

E4 = Leu–Met–Ala (residues 34 to 36)

E1 = Ala (residue 33)

E6 = Glu–Met–Lys–(Asp,Glu)–Trp–Tyr–Gly (residues 25 to 32)

"From the chymotryptic sequence it is clear that tryptophan must be one of the first 10 residues. This is possible only if E7 is the *N*-terminal elastase fragment, in which case tryptophan must be one of the first eight residues. This in turn is possible only if CT4 precedes CT5. The chymotryptic sequence, then, is CT4–CT5–CT8–CT1–CT6–CT3–CT7–CT2, and residues 1 to 31 are (Arg,Glu$_2$)–Trp–(Ala,Arg,Glu,Ile,Ser)–Phe–(Ala,Arg$_2$,Asp$_3$,Cys,Glu$_2$,Gly,His,Ile$_2$,Leu$_2$)–Met–Lys–(Asp,Glu)–Trp–Tyr.

"This then establishes the tryptic sequence as T1–T5–T6–T4–T9–T8–T3–T2–T7. Again, the first 31 residues are Arg–Glu–Glu–Trp–Ile–Arg–(Ala,Glu,Ser)–Phe–Glu–Arg–(Asp,Cys,His,Leu)–Arg–(Ala,Asp$_2$,Glu,Gly,Ile$_2$,Leu)–Met–Lys–(Asp,Glu)–Trp–Tyr

"The elastase sequence is E7–(E2,E9)–E6–E1–E4–E5–E8–E3. Fragment E7 = Arg–Glu–Glu–Trp–Ile–Arg–(Glu,Ser)–Ala (residues 1 to 9). This is followed by residues 10 to 23 in fragment E9 = Phe–Glu–Arg–(Asp,Cys,His,Leu)–Arg–(Asp$_2$,Ile$_2$,Leu)–Ala."

"Wait a minute," Michael interrupted. "Don't you know the amino acid next to the *C*-terminus of E9 from the carboxypeptidase/lithium borohydride experiment?"

"That's very true," answered Rayna. "Do you know its identity?"

"Have a heart," said Michael. "I'm beginning to understand the logic, but I still have no clue about the actual chemistry."

"Stick with this for a while longer and you'll begin to pick it up," she replied. "Position 22 turns out to be asparagine; thus E9 = Phe–Glu–Arg–(Asp,Cys,His,Leu)–Arg–(Asp,Ile$_2$,Leu)–Asn–Ala. E2 = Gly follows E9, and the entire sequence is:

E7–E9–E2–E6–E1–E4–E5–E8–E3 =
Arg–Glu–Glu–Trp–Ile–Arg–(Glu,Ser)–Ala–Phe–Glu–Arg–(Asp,Cys,His,Leu)–
Arg–(Asp,Ile$_2$,Leu)–Asn–Ala–Gly–Glu–Met–Lys–(Asp,Glu)–Trp–Tyr–Gly–
Ala–Leu–Met–Ala–Arg–Thr–Trp–Ala–Arg–Asp–Tyr–Asp–Arg–(Asp$_2$,Glu,Ile)–
Met–Ala–Phe–Thr

"Finally, let's look at the peptic fragments. We'll be able to place every one of them in the peptide:

P5 = Arg (*N*-terminal)
P6 = Glu–Gln (residues 2 and 3)
P8 = Trp–Ile–Arg (residues 4 to 6)
P11 = Glu–Ser–Ala (residues 7 to 9)
P2 = Phe (residue 10)
P18 = Glu–Arg–(Asn,Cys,His) (residues 11 to 15)
P16 = Leu–Arg–(Asn,Ile) (residues 16 to 19)
P19 = Leu–Ile–Asn–Ala–Gly (residues 20 to 24)
P15 = Glu–Met–Lys–Asn (residues 25 to 28)
P3 = Glu (residue 29)
P1 = Trp (residue 30)
P12 = Tyr–Gly–Ala (residues 31 to 33)
P17 = Leu–Met–Ala–Arg–Thr (residues 34 to 38)
P9 = Trp–Ala–Arg (residues 39 to 41)
P4 = Asp (residue 42)
P13 = Tyr–Asn–Arg (residues 43 to 45)
P14 = (Asp,Glu,Ile) (residues 46 to 48)
P10 = Asp–Met–Ala (residues 49 to 51)
P7 = Phe–Thr (residues 52 to 53)

"When P18 was treated with amidase, asparagine was converted to aspartate. Redigestion with pepsin cleaved at the new aspartate residue, showing that the original sequence was Glu–Arg–Asn–(Cys,His). Then the carboxypeptidase/lithium borohydride treatment yielded Glu–Arg–Asn–Cys–His (residues 11 to 15). Treatment of P16 with carboxypeptidase/lithium borohydride yields Leu–Arg–Ile–Asn (residues 16 to 19).

"Only two fragments remain to be placed. P14, a tripeptide, contains residues 46 to 48, Glu–Ile–Asn (notice that here we used the lithium borohydride data), and P19, a pentapeptide, contains residues 20 to 24, Leu–Ile–Asn–Ala–Gly. So the entire peptic sequence is

P5–P6–P8–P11–P2–P18–P16–P19–P15–P3–P1–P12–P17–P9–P4–P13–P14–P10–P7 =
Arg–Glu–Gln–Trp–Ile–Arg–Glu–Ser–Ala–Phe–Glu–Arg–Asn–Cys–His–Leu–Arg–Ile–Asn–Leu–Ile–Asn–Ala–Gly–Glu–Met–Lys–Asn–Glu–Trp–Tyr–Gly–Ala–Leu–Met–Ala–Arg–Thr–Trp–Ala–Arg–Asp–Tyr–Asn–Arg–Glu–Ile–Asn–Asp–Met–Ala–Phe–Thr

"Converting to the single letter code gives REQWIRESAFERNCHLRINLINAGEMKNEWYGALMARTWARDYNREINDMAFT.

"Here's another map showing all the cleavage sites." Rayna showed them the second diagram (Figure 24).

Sam and Michael stared at the two cleavage maps.

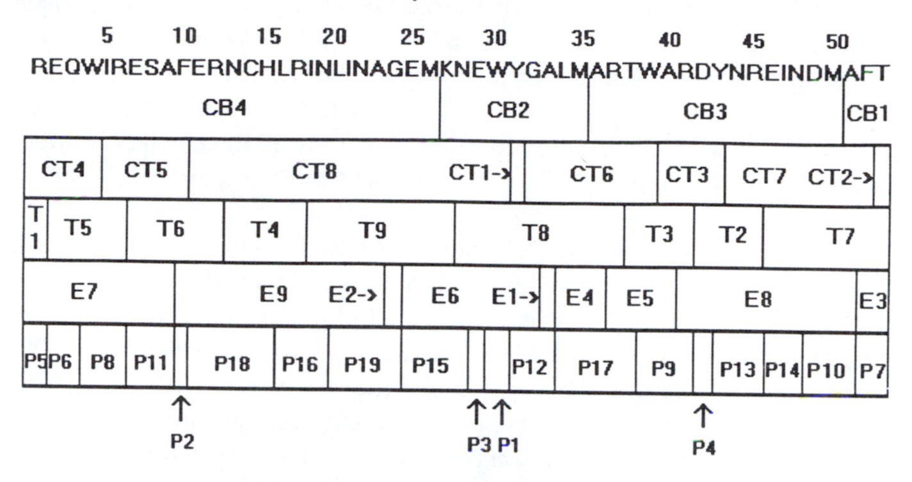

FIGURE 24 Cleavage map for peptide B.

"I must say", Michael began, "I really am impressed by the way you two solved these things. It's a real mental workout, but it's quite fascinating. I suppose there are lots of rules and regulations regarding cleavage specificities."

"Actually," said Sam, "there are only a few important rules. The best part is when you know the rules and have a chance to apply them. It certainly is good exercise for the brain."

Michael countered, "But now that we've solved the sequences, what does it all mean?"

Michael's use of the collective pronoun was not lost on either Rayna or Sam.

Sam replied, "What you haven't seen is that someone is hiding messages in these sequences. Look carefully. Can't you see some words in there?"

"Oh!" said Michael. "Sure. But so what? There are random words here and there, but do they mean anything?"

"Beats me," Sam said.

"That's the last piece of the puzzle," Rayna interjected. "These two sequences are meaningless without understanding that they contain exons and introns. The numbers on that funny Exxon note provided the remaining data.

"Each digit refers to the number of contiguous letters we have to group from each peptide. It's not too difficult to decide where to start. Let's lay out the two sequences again."

A: LINKSDATETREADSKLEMRINGWANTMATFEARPAYMYLASER-ICESTRINK

B: REQWIRESAFERNCHLRINLINAGEMKNEWYGALMARTWARD-YNREINDMAFT

"And here are the 'telephone' numbers"

313–164–2213
242–422–5122(4)

"Let's start with the first number, the one beginning with 313, and alternate selections with the 242 number. So we'll take the first three letters of A, then skip the first three letters of B and keep the next two. Skip the next two letters of A (4 and 5) and keep the next one (letter 6), skip one letter (number 5) of B and keep the next four (6 through 9), and so forth. We can write all these letters in upper case, calling them Group I. At the same time, let's put the remaining letters in lower case, in Group II."

A: LINksDatetREAdsKlEMRINGwaNTMAtfEArpaymYLaSerICEstrink
B: reqWIrESAFernCHlRinlinaGEmkneWYgaLMARTwaRdYNreiNDMAFT

"There are clearly two problems here. First of all, of course, the letters we grouped make no sense. After all, what can we make of either LINWIDESAFREACH...etc., or reqksratetern...etc. Second, when we start with the 313 number we end up with four letters in each peptide which are anomalous. If we are to place in Group I the N and D at positions 48 and 49 of peptide B, as well as M, A, F, and T at positions 50 to 53, why not just combine them into a single digit? Similarly for peptide A. If we're supposed to place the s and t at 48 and 49, and the r, i, n, and k at 50 to 53, in Group II, why not combine them? This must mean that we should start with the 242 number, rather than the 313 number.

"Now if we repeat the process, again listing the Group I letters in upper case and the Group II letters in lower case, look at what we get."

A: LInksDATEtREadsKlEMringwaNTmatfEARPAymYlaSErICestRINK
B: reQWIresaFerNCHlRinLINAGEmkNEWYgalmaRTwARdyNreINDmaft

"Group II, in lower case — renksresaterads — again, makes no sense. But look at what comes out in Group I."

LIQWIDATEFRENCHKREMLINAGENTNEWYEARPARTYARSENICINDRINK

Michael and Sam stared intently, comprehension slowly contorting their faces. Michael spoke first.

"'Liquidate French Kremlin agent? New Year party. Arsenic in drink!' Wow! If I hadn't seen you do this, step by step, I'd never have believed it. It's just incredible that anyone could be clever enough to think of coding a message this way, and even more amazing that you were able to decode it. What will you do now?"

"Nothing immediately," Rayna answered. "Oh, tomorrow morning I'll bring it to the attention of the Director. He won't understand the biochemistry, of course, but I'm sure he'll appreciate the gravity of the message."

"For now," she said, her eyes twinkling, "how about a little celebration? I think I have a bottle of fairly decent wine just waiting for the right occasion."

As they drank and further discussed their solution, Michael suddenly brightened.

"You know," he began. "What we've done here might have some greater significance. Suppose things like this happened in real life. I mean, if nature could extract sections of DNA and rearrange them, it might have all kinds of implications for evolution. And suppose you could do this in the lab. That would be an incredibly powerful tool for moving genes from one organism to another."

"Ah Michael," Rayna sighed, "I'm afraid you're just a few years late. Recombinant DNA technology has already been invented, and it's being used in labs everywhere. But I am impressed that you're understanding the logic of all this."

Michael smiled, mentally patting himself on the back and beginning to get the feeling that one needn't be a scientist to appreciate the beauty of biochemistry.

Summertime, and Is the Living Easy?

"It's definitely what they call an 'inside the Beltway problem,' Sam. No one outside Washington could possibly appreciate summer weather here."

Sam looked up from his salad, crunched a carrot stick in his mouth, and gave Rayna a quizzical look.

"I think the expression 'inside the Beltway' is usually reserved for political diatribes," he said. "Certainly the newspapers use it that way."

"That may be," Rayna replied, "but on a deeper level I think that it can apply to anything that's peculiarly Washingtonian. And the weather here is certainly peculiar."

Finishing the last of her strawberry-banana yogurt, she began expounding on the dubious joys of Washington in summer. Jabbing her plastic spoon in the air for emphasis, she continued, fulminating with mock seriousness until Sam, unable to keep a straight face, laughed out loud.

"Despite all the kidding, Sam, I'm serious about this summer heat. Sultry is probably the best way to describe it. We're fortunate enough to work in an air conditioned office, but do you remember the time the air conditioning broke down?"

"Who could forget?" Sam replied. "It was like one of those legendary Turkish baths. But you're right about the summer. Between the heat and the humidity, sometimes I just feel as if I'm melting into the sidewalk.

"And speaking of being outdoors," he continued, "we still have a date for tomorrow, don't we?"

"I'm looking forward to it. You know, there's nothing like a good old July Fourth picnic and fireworks on the Mall to rekindle sagging patriotism. Oh, look at the time. We'd better get back to work, or we won't have jobs to pay for vacations."

As they disposed of the remains of their lunch, Rayna caught sight of a faint movement in a corner of the cafeteria. Allowing Sam to lead the way out, she surreptitiously took a quick side step into a dimly lit alcove where a shadowy figure handed her a small packet of papers. The whole transaction required no more than five seconds, and Sam never knew that anything had occurred.

Back in her office Rayna swiftly shuffled the new papers in with a few file folders on her desktop, then extracted them again and quite ingenuously said to Sam, "Oh, look at this... more gels." The picture she pulled out is shown in Figure 25.

"And here is a list of restriction enzymes responsible for fragmentation." The list is shown in Table 8.

"Sam, how would you like to help decipher these?" she asked with mock sweetness. "Maybe there's another secret message."

Sam, recalling the heartstopping excitement generated by the Kennedy Center gels, agreed despite his reluctance to get involved in another adventure. He took the gel photographs and the list of enzymes from her and, disappearing into his own cubicle, collected a magnifying glass, ruler, and pencil and began to measure migratory distances. Some time later, looking slightly bedraggled and somewhat more bleary-eyed, he returned.

"I hate this," he began. "I don't want to sound obnoxious, but if I don't go blind measuring these things I'll surely die of boredom. I've measured all the distances, and what a pain it's been, but I can't calculate any fragment lengths without knowing what the standards are."

"You're absolutely right, Sam," she replied. "My guess is that this is probably the same collection of 1 Kb Ladder fragments that we saw last time. Look, I have a spreadsheet in the computer. All you have to do is enter the measured distances and it will give you corresponding fragment lengths."

"That would be a great help," he said, sitting down in front of the monitor. "But what would be even better would be a program to read these gels."

It took Sam no more than a half-hour to coax a complete listing of fragment lengths from the computer.

"That's certainly a lot faster than plotting everything by hand," he said, grinning. "It's fortunate that you had that regression template to do the calculations. But look at what I found. If you sum the fragment lengths in each digest you get exactly 3.0 kbp for almost every one of them. But a few are off, by as little

Table 8

Identities of Restriction Enzymes for Gels A to D

GelA	GelB	GelC	GelD
Afl II	Nco I	Afl II/Nbl I	BseP I/Nde I
Asn I	Nde I	Asn I/BamK I	Hpa I/Sfu I
BamK I	Pvu II	Asn I/EcoR V	Nbl I/Nco I
BseP I	Sac I	Asn I/Sac I	Nbl I/Nde I
EcoR V	Sca I	Asn I/Sfu I	Nbl I/Sac I
Fsp I	Sfu I	Asn I/Stu I	Nco I/Sca I
Hpa I	Spa I	BamK I/Spa I	Nde I/Stu I
Nbl I	Stu I	BseP I/Fsp I	Pvu II/Spa I

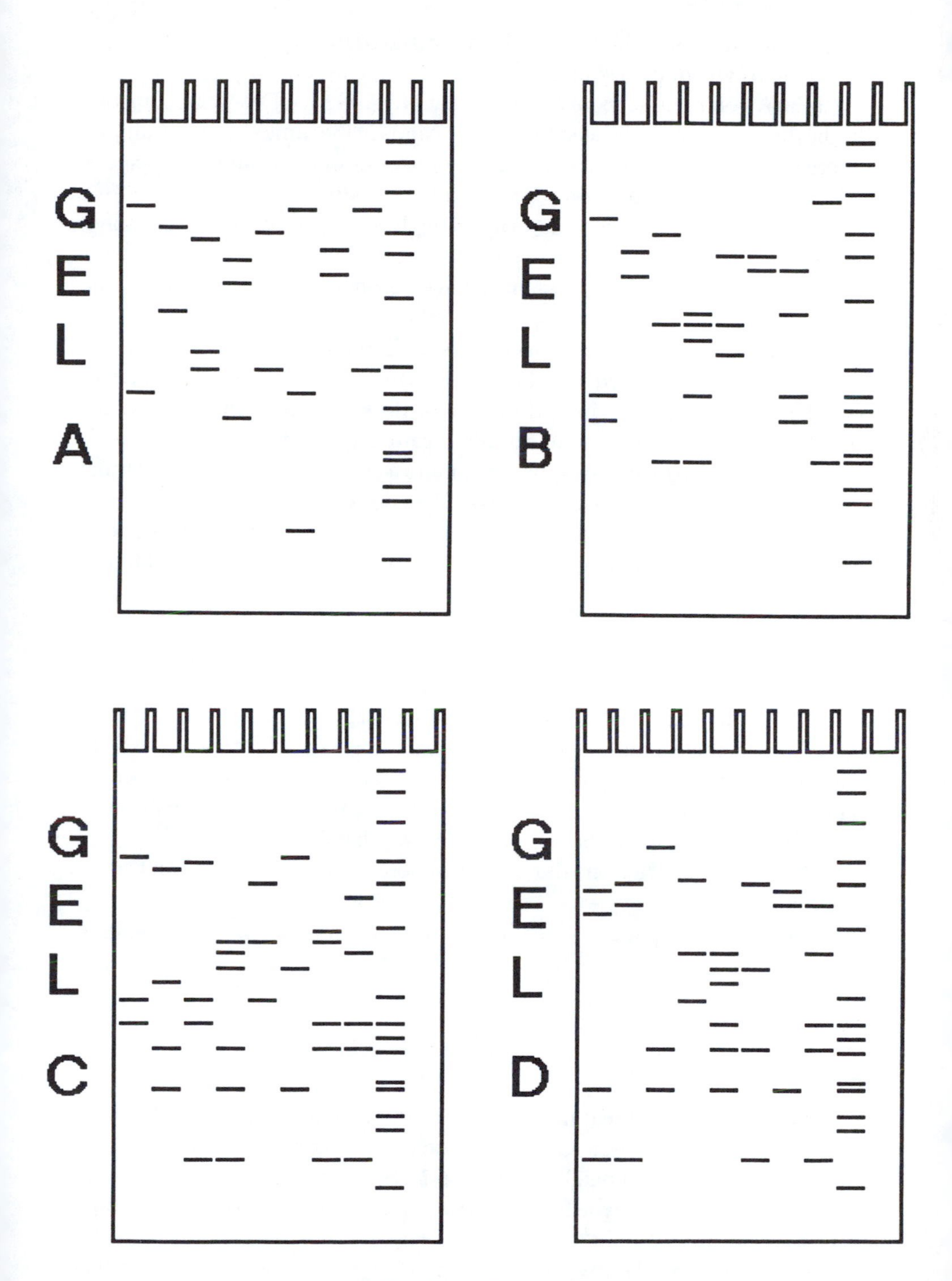

FIGURE 25 Gels A to D.

as 0.1 or as much as 0.5 kbp. What do you make of that?"

Rayna frowned, then smiled.

"I think you're right about the molecular weight of the DNA, Sam. It looks like the discrepancies are related to multiple bands of the same size. For example, if there were two bands of 0.2 kbp, we'd see only one of them, and the apparent molecular weight would be only 2.8 instead of 3.0 kbp."

"That sounds reasonable," Sam said. "I might as well go and see if I can solve the restriction map now."

"Good," said Rayna. "You seem to have become pretty confident of your abilities."

"Practice," he mumbled sarcastically, shuffling out of her office.

By the time he returned with his results, Sam's dismay clearly had intensified. Slamming a sheaf of papers down on the desk he said, "Here it is. I've got the whole bloody map for you, but it's meaningless. I even converted each restriction sequence to its corresponding amino acid, just as we did the last time, and you know what came out? Garbage! Here, look."

Rayna picked up the top page of the pile and read the results of Sam's labors.

"I can't make anything of it," said Sam. "I can see 'LID' and 'RATS' but not much else."

"You know, Sam, I wonder if this message might not be a little more sophisticated than the last one. Last time each restriction site was read out so that the triplet on the 5′-end coded for the *N*-terminal amino acid of a pair, even though the individual DNA codons were read from 3′ to 5′. Suppose they finally figured out their error and read the recognition sites themselves from 3′ to 5′. In that case each pair of letters would be reversed. Does this make any more sense? There's 'CAR', 'DIES', 'ACE', 'DIESEL', 'FEAR', 'MARE', 'CAP'. But so what?"

"I'm totally stumped," Sam replied. "Now what do we do?"

"Well," she said, "for one thing, I have another piece of data. I didn't show it to you earlier, because I wasn't sure we'd need it. Look at this."

A small sheet of paper, evidently torn from a spiral-bound notebook, bore the inscription

AD	A5
E6	E5
E5	6
9	FA

"And what does all that mean?" Sam said. "Wait a minute. I bet it's some kind of musical progression. Chords, or some such thing. Why don't you take it home to that musician neighbor of yours? I bet he can figure it out."

"Good idea, Sam. Maybe by the time you pick me up tomorrow we'll know what we're facing."

But by the next day, when Sam arrived at Rayna's apartment, matters appeared no better.

"Oh Sam," she began morosely. "I'm stumped. Michael looked at that list last night. He must have spent hours picking out notes, juggling chords, inverting, modulating, who knows what else. Poor guy. He said nothing made any sense musically. It didn't refer to any music he knew; there were no recognizable melodies — nothing. He even did a computer search to try to match what I gave him to previously published music. No luck. I'm afraid your idea didn't help much."

Assuming a lighter demeanor, Sam tried to put everything in perspective.

"Okay," he said. "Enough business for now. This is a holiday; the government has given us permission to leave our work at the office. Let's head downtown for the Fourth."

Trying to avoid demolishing Sam's ebullience, Rayna managed a fragile smile as she followed him out of the apartment.

The National Mall lies on a enormous expanse of prime downtown Washington real estate, its grassy open space stretching from the domed Capitol building at the east end to the obelisk of the Washington Monument at the west and spanning the equivalent of four city blocks from Constitution Avenue on the north to Independence Avenue on the south. Both sides are lined with what is probably the most concentrated collection of museums in the world: the circular Hirshhorn Museum, with its outdoor sculpture garden and peculiar, eccentric fountain; the red brick Arts and Industries Building, containing a re-creation of the Victorian era Centennial Exhibition; subterranean galleries of Asian and African art; and the National Air and Space Museum, sporting a special outdoor display of small aircraft, all on the southern edge; and on the northern, the National Gallery of Art, the National Museum of Natural History, and the National Museum of American History. In the midst of the entire complex, standing as the overseer of all activities, the Smithsonian Castle, headquarters of the vast Smithsonian Institution empire, with its information desk brimming with descriptive multilingual brochures, serves as the starting point for the visits of countless tourists.

As Rayna and Sam emerged from the ultramodern Metro subway they found themselves dazzled by the midsummer sunlight reflecting off the bright green grass of the Mall. The crowds had not yet begun to gather in earnest, but they knew from experience that within a few hours the Mall would be a sea of humanity.

"Sam, it's still early, and we don't want to spend too many hours baking in the sun. How about a quick run through one of the museums."

"Sure," he said. "I'd really like to look at the new computer exhibit in the American History museum."

Crossing to the north side of the Mall, they climbed a short flight of granite steps passing under *Infinity*, an abstract stainless steel sculpture rotating gracefully atop its pedestal, and entered the museum. After pausing to adapt to the darkness they quickly made their way to the new exhibit, where starry eyed children, and not a few similarly entranced adults, engaged in hands-on exploration of do-loops, feedback controls, and robotics. As they separated to circle a display in opposite directions, Rayna peered closely at a board on which lights of various colors flickered on and off in apparently random fashion. This demonstration of computer memory utilization triggered a vague recollection that she couldn't quite raise to her consciousness, and forced her into dark and distracted introspection as she struggled to make contact with a lost thought.

Sam, arriving from the other side of the display, smiled broadly.

"It's pretty sexy in here, isn't it. Look at those kids," he exclaimed. "This stuff is music to their ears. They're eating it up, if you'll pardon a mixed metaphor."

"What?" she said absently.

"I said..." he began.

Suddenly Rayna turned to him.

"Of course," she cried. "It's not music at all. Sam, we've got to get over to the castle. I just hope they have a terminal we can borrow. Hurry!"

Having grown accustomed to Rayna's sudden bursts of insight, Sam had learned not to bother with questions when she experienced what he liked to call "the light bulb phenomenon." He followed her outside and across the Mall, weaving through the accumulating mass of holiday revelers. Inside the castle Rayna identified herself to the clerk on duty and got directions to the computer room in the basement. Once there she appropriated a terminal for herself and logged in to the agency's mainframe.

A few rapidly keyed in commands got Rayna all the information she needed. She logged off the computer and turned to Sam.

"It's just as I thought, Sam. These are hexadecimal digits. If I'm right, this can solve the message in the gels."

After a few minutes of scribbling Rayna put down her pencil, turning her ashen face to Sam.

"We've got to get to the Archives as soon as possible," she said, snapping the switch to turn off the terminal. "Meet me upstairs at the main entrance. I'll be back in a few minutes. There's one more thing I have to do."

Presently she charged up the stairs, taking two at a time, and rejoined Sam while stuffing a packet into her purse. Together they made their way once again outside.

The National Archives, on the north side of the Mall and a couple of blocks east of the castle, serves as the national repository for documents and artifacts. The original building, of course, had rapidly become overburdened, necessitating the construction of a number of remote auxiliary facilities. Nonetheless the showcase of the collection, the original Declaration of Independence and Constitution, remained on display here, in hermetically sealed, light-filtering cases, under helium atmosphere, the former displayed vertically just above eye level and the latter, along with the Bill of Rights, in a horizontal case just below. The documents were irreplaceable, and an elaborate security system had been designed for their protection. Each night at closing time the cases descended into a 7 by 5 by 6 foot vault surrounded by a shell of reinforced concrete and steel, some 22 feet below the exhibition hall.

On a normal day the walk from the castle to the Archives would take no more than five or six minutes. Today, however, with some quarter of a million people gathering on the Mall for a day of picnicking to be followed by a spectacular fireworks display in celebration of the national holiday, the going would be much slower. Families and extended families had spread picnic blankets on the grass, babies slept on the ground, and most of the remaining space had been appropriated by groups of self-styled athletes engaged in barely organized games of Frisbee®, football, volleyball, and softball, as well as others of more obscure origin and even less-discernible rules. Observing the entire scene, but maintaining a low profile, officers of the Metropolitan Police and the National Parks Service stood ready to intervene in any potentially explosive situation. Hence Rayna and Sam, in order to avoid raising any suspicions, found it prudent to make their way across the Mall with agonizing slowness, aware that any disturbance they might cause could easily doom their mission.

Finally attaining their goal, they bounded up the broad concrete stairway leading into the Archives building and entered the cool Rotunda, whose muted elegance seemed to constrain visitors to speak in whispers. Rayna rapidly took in the scene, noting the armed guards protecting the display cases which held the precious documents. Nothing seemed amiss. Forward of the centerpiece and hugging the walls to the left and right, curved rows of glass cases held artifacts and displayed documents describing the history and chronology of the Declaration and Constitution. A line of onlookers snaked past these cases, some taking the time to read carefully, others apparently interested only in telling the folks back home that they had actually seen the foundations of their government.

Behind the Rotunda itself, the Circular Gallery (actually more of a semicircle) approached from the left and right sides of the main entrance and surrounded the central exhibition hall.

"Wait here, Sam," Rayna said.

Sam queried her with raised eyebrows.

"I've got to use the ladies' room," she said abashedly.

During Rayna's brief absence, Sam tried to look nonchalant while gazing resolutely at each of the tourists. Not knowing what he was searching for, Sam found it especially difficult to try to single out anyone who might look suspicious. In fact, feeling all the more conspicuous for standing about idly rather than passing along the displays as any normal tourist would do, he decided to venture off to the side to inspect the Circular Gallery. As he reached its most remote point, directly behind the documents and separated from them by a single wall, he realized that he stood in the only position which allowed him to observe events in both wings of the Gallery simultaneously.

To his left and to his right two men methodically knelt to remove objects from their backpacks. Terror-stricken, Sam stared first at one and then the other.

Suddenly he realized that they were assembling assault weapons. Desperately trying to avoid both panic and notice, he made his way back to the Rotunda rapidly searching out Rayna. Swiftly yet quietly, she returned to his side, her eyes darting animatedly from one side to the other.

"Watch the exit, Sam," she whispered.

Sam looked at her curiously, as a woman came running into the Rotunda from a door on the right side. Suddenly he heard a vague crackling as if distant gunshots had penetrated his consciousness. It's too early for the fireworks to start, he thought incongruously.

"Fire! Fire! There's a fire in the ladies' room!" the woman yelled.

The security guards immediately leaped to attention, just as the steel doors on the vertical display case snapped shut and the hydraulic rods lowered the documents to safety beneath the floor. Sam, looking through the panic-stricken crowd, tried to spot the two men with backpacks, finally locating them just as they ran out the doors to dissolve into the crowd outside. Inside, the guards tried to maintain some semblance of order, assuring the assembled multitude that the fire was contained, that the automatic sprinkler system had been activated, and that the danger had passed. As a precaution, however, they announced that the hall would have to be evacuated.

Outside, once again in the bright sunshine, Sam squinted at Rayna.

"You didn't have anything to do with that fire, did you?" he asked.

"Who, me?" she replied guilelessly. "Would I do something like that? There was a trash barrel filled with paper towels. I guess it must have been a case of spontaneous combustion. Did you see anything suspicious while I was away?"

"Well, there were two guys who seemed to be trying to get something out of their backpacks, but they disappeared as soon as that woman came running out. I don't suppose we'll ever know who they are."

"Probably not," said Rayna. "In any case, the assault never occurred, and in all probability the public will never know what really happened."

"What about those gunshots I heard?" Sam continued.

"Just a little precaution to be sure the guards were awake," she grinned. "Do you remember that chemical magic show I used to do at our Christmas parties? One of the tricks involved minor explosions as people walked across the floor. It's really quite simple. I soak some filter paper in a solution of inorganic salts and then dry it out. Before we left the castle I made a quick stop into the preservation studio. They had everything I needed. I just dropped a few shreds of prepared filter paper outside the door to the ladies' room. That poor panic-stricken woman who discovered the fire did the rest."

"I suppose you're going to tell me how you finally figured out the message," Sam said hopefully.

"Oh, eventually," Rayna teased. "But let's leave that for later. I'm hungry, and the museum restaurants will probably be closing soon. And later I'd like to see some real fireworks."

Solution

As the last of the fireworks exploded behind the Washington Monument, the quarter million onlookers applauded vigorously and in a spontaneous outpouring of emotion, began singing "Happy Birthday, America" over and over again. Finally they began to disperse, the majority heading for the subway entrances. Sam and Rayna found a bench along one of the Mall walkways and decided to allow the crowd to thin out before attempting to leave.

"Well, I assume you've had your fill of fireworks," he said. "Now how about telling me how you solved that message."

"Good idea," she responded lightheartedly. "What does it matter if a zillion people overhear?

"Here is your computer printout listing the calculated fragment lengths. As you pointed out, with the exception of one single digest and four double digests, everything else is nice and neat, with each digest showing exactly 3.0 kbp for the length of the DNA. The EcoR V digest shows two bands totaling 2.5 kbp. The only thing that makes any sense in this context is a second 0.5 kbp band. The same thing is true for the atypical double digests: both BamK I/Spa I and Nco I/Sca I have a second 0.3 kbp band, Nbl I/Nco I has a second 0.2 kbp band, and Pvu II/Spa I has a second band of 0.1 kbp. It's only when you recognize this that you can solve the restriction map."

The data are summarized in Table 9.

"Your map was quite correct (Figure 26), and it didn't take you too long to compile the list of amino acids in the correct order and to convert them to their one letter codes. The message that came out was ACPRCAKLIDSE-CALENIIDSELENVEFRATSMHRALELQCALQWPTSSRACWPLE. The only problem was that it made no sense. I was right, by the way, when I suggested that the recognition sites were actually read correctly, from 3' to 5', but I only discovered that later. And even if we had known that at the time, the message would have been CARPACLKDIESACELINDIESELVNFEARSTHMARE-LQLACQLPWSTRSCAPWEL, which still told us nothing."

Sam interrupted the flow of Rayna's story.

"That little piece of paper with the funny inscription. The one that we thought was some kind of musical notation. Where did that come from?"

"I could tell you, Sam," Rayna said with a diabolical smile. "But then I'd have to kill you. Seriously, though, there are some sources that I'm afraid will have to remain undisclosed.

"I tried to work with Michael on deciphering that last bit of code in terms of music, but it turned out to be a red herring. By the time we left for the Mall, I really had no idea what we were facing."

"But the computer exhibit at the museum helped?" Sam inquired.

"That, and your offhand remark about the kids looking at it."

Sam looked puzzled.

"What did I say? Something about how they were enjoying it, and that it was music to their ears?"

Table 9

Calculated Sizes of Restriction Fragments (Gels A to D)[a]

	Gel A		Gel B		Gel C		Gel D	
	Distance*	**kb**	**Distance**	**kb**	**Distance**	**kb**	**Distance**	**kb**
Lane 1	**Afl II**		**Nco I**		**Afl II/Nbl I**		**BseP I/Nde I**	
	12.4	2.6	14.2	2.3	15.8	2.1	20.6	1.5
	40.2	0.4	40.2	0.4	36.4	0.5	24.1	1.2
			43.9	0.3	40.2	0.4	50.0	0.2
							60.3	0.1
	Sum:	3.0		3.0		3.0		3.0
Lane 2	**Asn I**		**Nde I**		**Asn I/BamK I**		**Hpa I/**	**Sfu I**
	15.7	2.1	19.0	1.7	17.6	1.9	19.7	1.6
	28.2	0.9	22.6	1.3	34.1	0.6	22.8	1.3
					43.9	0.3	60.3	0.1
					49.9	0.2		
	Sum	3.0		3.0		3.0		3.0
Lane 3	**BamK I**		**Pvu II**		**Asn I/EcoR V**		**Nbl I/**	**Nco I**
	17.5	1.9	16.5	2.0	16.5	2.0	14.3	2.3
	34.1	0.6	29.8	0.8	36.9	0.5	44.0	0.3
	36.8	0.5	50.0	0.2	40.2	0.4	50.0	0.2
					60.3	0.1		
	Sum	3.0		3.0		3.0		2.8
Lane 4	**BseP I**		**Sac I**		**Asn I/Sac I**		**Nbl I/Nde I**	
	20.6	1.5	28.1	0.9	28.2	0.9	19.1	1.7
	24.1	1.2	29.8	0.8	29.9	0.8	29.9	0.8
	44.0	0.3	32.1	0.7	32.1	0.7	37.0	0.5
			40.2	0.4	44.0	0.3		
			50.0	0.2	49.9	0.2		
					60.3	0.1		
	Sum	3.0		3.0		3.0		3.0
Lane 5	**EcoR V**		**Sca I**		**Asn I/Sfu I**		**Nbl I/Sac I**	
	16.4	2.0	19.5	1.6	19.5	1.6	29.9	0.8
	36.9	0.5	29.8	0.8	28.2	0.9	32.1	0.7
			34.1	0.6	36.9	0.5	34.2	0.6
							40.4	0.4
							44.1	0.3
							50.2	0.2
	Sum	2.5		3.0		3.0		3.0
Lane 6	**Fsp I**		**Sfu I**		**Asn I/Stu I**		**Nco I/Sca I**	
	13.1	2.5	19.5	1.6	15.8	2.1	19.6	1.6
	40.2	0.4	21.6	1.4	32.2	0.7	32.1	0.7
	60.6	0.1			49.9	0.2	43.9	0.3
							60.3	0.1
	Sum	3.0		3.0		3.0		2.7

Table 9 (continued)

Calculated Sizes of Restriction Fragments (Gels A to D)[a]

	Gel A		Gel B		Gel C		Gel D	
	Distance*	kb	Distance	kb	Distance	kb	Distance	kb
Lane 7	Hpa I		Spa I		Bamk I/Spa I		Nde I/Stu I	
	19.1	1.7	21.6	1.4	26.8	1.0	20.6	1.5
	22.7	1.3	28.2	0.9	28.2	0.9	22.8	1.3
			40.1	0.4	40.2	0.4	50.0	0.2
			43.9	0.3	43.9	0.3		
					60.3	0.1		
	Sum	3.0		3.0		2.7		3.0
Lane 8	Nbl I		Stu I		BseP I/Fsp I		Pvu II/Spa I	
	13.0	2.5	11.4	2.8	21.7	1.4	22.8	1.3
	36.8	0.5	50.0	0.2	29.9	0.8	29.9	0.8
					40.2	0.4	40.3	0.4
					43.9	0.3	44.0	0.3
					60.3	0.1	60.3	0.1
	Sum	3.0		3.0		3.0		2.9

[a] Slope = –33.7, intercept = 128 (for each gel).

* All distances are in millimeters.

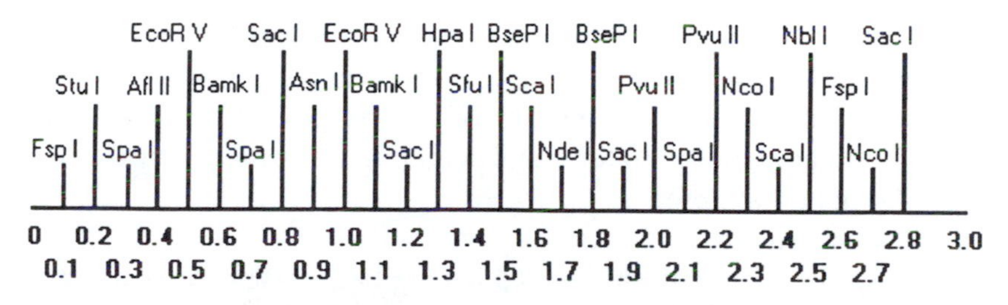

FIGURE 26 Restriction map for Archive gels.

"Actually you said the exhibit was sexy. I guess that was the prod I needed to make the mental connection. Not to 'sex', but to 'hex'. The notations weren't music, but hexadecimal digits. Do you remember when we first took that computer course and the instructor explained the binary system of notation, and showed how long strings of zeros and ones could be represented conventionally in decimal notation, but that it was even more convenient to use hexadecimal? Normally a list of hexadecimal digits is preceded by a dollar sign to distinguish them from ordinary decimal values. My source didn't do that, so it took a little more time to understand what we had.

"But once I realized that, I was able to get the computer to translate the hex digits back to binary. It turned out that the number of binary digits was exactly the same as the number of letters in the decoded restriction map. I lined up the two lists and assumed, correctly, that a '1' indicated a letter to keep while a '0' indicated a letter to drop.

"Here are the hexadecimal-binary interconversions: $5 = 0101, $6 = 0110, $9 = 1001, $A = 1010, $D = 1101, $E = 1110, and $F = 1111. Then when I lined up the text with the appropriate binary digits this is what I got."

A	D	A	5	E	6	E
1 0 1 0	1 1 0 1	1 0 1 0	0 1 0 1	1 1 1 0	0 1 1 0	1 1 1 0
A C P R	C A K L	I D S E	C A L E	N I I D	S E L E	N V E F
C A R P	A C L K	D I E S	A C E L	I N D I	E S E L	V N F E

5	E	5	6	9	F	A
0 1 0 1	1 1 1 0	0 1 0 1	0 1 1 0	1 0 0 1	1 1 1 1	1 0 1 0
R A T S	M H R A	L E L Q	C A L Q	W P T S	S R A C	W P L E
A R S T	H M A R	E L Q L	A C Q L	P W S T	R S C A	P W E L

"The first message gives APCALISAENIIELNVEASMHREQALWSS-RACWL, which was clearly gibberish. But the second one gives CRACKDECLI-NDSEVNFRTHMALLCQPTRSCAPE.

"I see some of it," Sam waxed enthusiastic. "'CRACK DECL IND'. That points to the Declaration of Independence. 'SEVN FRTH'. Seven fourth. Oh, the fourth of July. 'MALL'. 'CQPTR'. The helicopters that were on display outside the Air and Space Museum. One of them was the getaway. It's cute of them to use a 'Q' in place of an 'O'. And the last word, 'SCAPE'. Escape! Crack the Declaration of Independence on July fourth. Escape using one of the helicopters parked on the Mall. Wow! Do you think anyone will believe this?"

"Frankly, I'm not sure we should reveal it to anyone," Rayna said darkly. "It's sufficient that we thwarted the plot. I don't think anyone will try this again. Why get the public all upset?"

"I see what you mean," he agreed.

The crowd of celebrants on the Mall having largely dissipated, a peaceful silence supplanted the frenzy of a few hours earlier. To the east, illuminated by pale fluorescent light, the Capitol dome sat in ghostly silence, while at the opposite end of the Mall the Washington Monument, its red aircraft warning lights alternately blinking, stood guard over the federal city. North of the monument, in the White House, the President prepared for another day in the seat of power, unaware of the high drama of the afternoon. As a warm breeze blew in off the Potomac River, tossing about the discarded remnants of thousands of carefree picnics, Rayna and Sam crossed the Mall in somber silence. Walking through the darkness to catch the last Metro train out of downtown Washington, each speculated on the potentially calamitous consequences of the day's events.

Appendix A Solutions to End of Chapter Problems

Chapter 1

Problem 1.1

A. (Ala,Cys,His_2,Leu_2,Thr,Val)

B. (Arg,Asp_2,Leu,Lys,Phe,Pro,Val_2) — Note that Asn is hydrolyzed to Asp.

C. (Ala_2,Glu,His_2,Leu,Lys_2,Pro,Ser,Thr)

D. (Ala,Asp,Glu,Ile,Lys,Ser,Thr) — Note that Gln and Asn are hydrolyzed to Glu and Asp, respectively.

E. (Arg_2,Asp_4,Cys_3,Gly_2,Ile,Leu,Pro_2,Ser_2,Thr,Trp) — It is assumed that the disulfide is reduced prior to hydrolysis.

F. (Ala,Arg_4,Asp_2,Glu_3,Gly_2,His,Leu_6,Phe,Ser_4,Thr_2,Val) — Note that the amide group at the *C*-terminus is hydrolyzed, just as the amide groups of Asn or Gln are; it has no effect on the amino acid analysis.

Problem 1.2

Asp–Asp → 2-amino-4-hydroxybutanoic acid + 2-amino-1,4-butanediol

Asp–Asn → 2-amino-4-hydroxybutanoic acid + 3-amino-4-hydroxybutanoic acid

Asn–Asp → aspartic acid + 2-amino-1,4-butanediol

Asn–Asn → aspartic acid + 3-amino-4-hydroxybutanoic acid

Problem 1.3

A. DNP–Ala + Glu + Pro + Ser + Tyr + Val + 2-aminoethanol

B. DNP–Ser + Ala + Arg + 2 Asp + Cys + Glu + His + Lys + 2-amino-3-phenyl-1-propanol

C. DNP–Asp + 3 Ala + Arg + Asp + Ile + Leu + Ser + Tyr + Val + 2-3-amino-3-mercapto-1-propanol

D. DNP–Lys + 2 Asp + 2 Glu + 2 Leu + 2 Lys + Tyr + 3-amino-4-hydroxybutanoic acid
E. DNP–Asp + Ala + Arg + Asp + Cys + Glu + Gly + Ile + Leu + Lys + Thr + 2-amino-1,5-pentanediol
F. DNP–Gly + 3 Asp + Cys + Ile + Lys + Val + Tyr + 2-amino-1,3-butanediol
G. DNP–Ser + Ala + Arg + 2 Asp + 2 Glu + Lys + Phe + Tyr + 2-amino-1,3-propanediol
H. DNP–Lys + Ala + 3 Arg + 2 Asp + Glu + 2 Gly + 2 Ile + Leu + Met + Pro + 4-amino-5-hydroxypentanoic acid
I. DNP–Ala + Ala + Asp + 2 Gly + His + Ile + 2 Lys + Phe + Ser + Thr + Val + 2-amino-4-methyl-1-pentanol
J. DNP–Leu + Ala + Asp + Gly + 2 His + Ile + Lys + Phe + Val + 2-amino-3-methyl-1-pentanol

Chapter 2

Problem 2.1

A. CNBr:
  Asp,Gly,Hsr,Lys,Pro,Thr (residues 1 to 6)
  Glu,Gly,Leu,Lys,Phe,Pro,Val (residues 7 to 13)

B. Trypsin:
  Gly,Lys,Pro,Thr (residues 1 to 4)
  Asp,Glu,Gly,Leu,Lys,Met,Phe,Pro,Val (residues 5 to 13)

C. Chymotrypsin:
  Glu,Gly,Leu,Lys,Pro (residues 9 to 13)
  Asp,Gly,Lys,Met,Phe,Pro,Thr,Val (residues 1 to 8)

D. Elastase:
  Gly,Pro (residues 1 and 2)
  Leu,Lys (residues 12 and 13)
  Glu,Gly,Phe,Pro (residues 8 to 11)
  Asp,Lys,Met,Thr,Val (residues 3 to 7)

E. Pepsin:
  Phe,Pro (residues 8 and 9)
  Glu,Gly (residues 10 and 11)
  Leu,Lys (residues 12 and 13)
  Asp,Gly,Lys,Met,Pro,Thr,Val (residues 1 to 7)

Problem 2.2

CB1, which contains no homoserine, must be the *C*-terminal peptide. Since the *C*-terminal amino acid is isoleucine, the sequence of CB1 is Val–Ile.

The FDNB experiment shows that the *N*-terminus of the heptapeptide is glycine. Since CB3 is the only fragment containing glycine, CB3 must be at the *N*-terminus of the heptapeptide. This allows placement of CB2 in the middle, so that the sequence can be shown as CB3–CB2–CB1.

The sequence of CB2 is Ala–Hsr, and the sequence of CB3 is Gly–Ile–Hsr. The sequence of the heptapeptide is Gly–Ile–Met–Ala–Met–Val–Ile.

Problem 2.3

Fragment T1 lacks basic amino acids and is therefore the *C*-terminal fragment. Since the lithium borohydride data show aspartate at the *C*-terminus of the heptapeptide, the sequence of T1 is Val–Asp.

From the FDNB data we conclude that serine is at the *N*-terminus. Since the only serine in the heptapeptide is in T3, we conclude that T3 is the *N*-terminal fragment and has the sequence Ser–Ala–Lys.

The only remaining fragment is T2 = Pro–Arg, which must be between T3 and T1. The entire sequence is T3–T2–T1 = Ser–Ala–Lys–Pro–Arg–Val–Asp.

Problem 2.4

Fragment CT2 is the *C*-terminal fragment. Since the FDNB data identify threonine as the *N*-terminal amino acid of CT2, we can show the sequence of CT2 as Thr–Gly–Gly.

We know, also from FDNB data, that the *N*-terminal amino acid of the nonapeptide is glycine. Although glycine appears in both fragments CT2 and CT3, only CT3 could be at the *N*-terminus since we have already located CT2 at the *C*-terminus. The sequence of CT3 is Gly–Val–Tyr. The remaining fragment, CT1, will be between CT2 and CT3; its sequence is Arg–Ser–Tyr.

Thus, the entire amino acid sequence of the nonapeptide is CT3–CT1–CT2 = Gly–Val–Tyr–Arg–Ser–Tyr–Thr–Gly–Gly.

Problem 2.5

The order of the elastase fragments is E3–E2–E1 = Ser–(Met,Pro,Tyr)–Ala–(Arg$_2$,Met)–Gly–Tyr. We know this because the FDNB data indicate *N*-terminal serine. E3 contains the only serine residue, and E1 contains no target amino acids, hence it is at the *C*-terminus. There is, however, no further information available from the elastase cleavage.

The cyanogen bromide data provide the remaining clues. Fragment CB1 must be at the *N*-terminus since it contains the only serine. The sequence of CB1 is Ser–Tyr–Met. CB2 is at the *C*-terminus. We already know from the elastase digest that the last two amino acids (positions 9 and 10) are Gly–Tyr. Therefore, the sequence of CB2 is Arg–Gly–Tyr. CB3 contains amino acid residues 4 to 7: CB1–CB3–CB2 = Ser–Tyr–Met–(Ala,Arg,Pro)–Met–Arg–Gly–Tyr.

Positions 1 through 3 are in CB1. E3 contains residues 1 to 5; we already know that position 5 is alanine, so that position 4 must be proline. We can show the sequence E3 = Ser–Tyr–Met–Pro–Ala. Since the sequence Pro–Ala corresponds to the first two positions of CB3 and the *C*-terminus of CB3 must be methionine, the only amino acid remaining (i.e., position 6 of the decapeptide) is arginine. The entire sequence is

E3–E2–E1 = CB1–CB3–CB2 = Ser–Tyr–Met–Pro–Ala–Arg–Met–Arg–Gly–Tyr

Problem 2.6

Fragment CB1, which lacks homoserine, is at the *C*-terminus. Since the *C*-terminal amino acid is glutamate, the sequence of CB1 is Ser–Glu. Fragment CB2 is a dipeptide; since homoserine is at its *C*-terminus, its sequence must be Val–Hsr (i.e., Val–Met).

The *N*-terminus of the original peptide is serine. Both CB1 and CB3 contain serine, but CB1 has already been established as the *C*-terminal fragment. Thus CB3 must be at the *N*-terminus and has the sequence Ser–Gly–Hsr (i.e., Ser–Gly–Met).

Fragment CB4 has homoserine (hence methionine) at its *C*-terminus, but the data do not allow assignment of the other two amino acids. Thus a partial sequence for CB4 is (Arg,Thr)–Met. It is also not known whether the overall sequence is

CB3–CB4–CB2–CB1 = Ser–Gly–Met–(Arg,Thr)–Met–Val–Met–Ser–Glu

or

CB3–CB2–CB4–CB1 = Ser–Gly–Met–Val–Met–(Arg,Thr)–Met–Ser–Glu

Tryptic fragment T2 lacks basic amino acids and is, therefore, at the *C*-terminus. Consequently, T1 is the *N*-terminal fragment. Since we already know the identities of residues 1 through 3, and the specificity of trypsin places arginine at position 4, this cleavage provides all the data needed to complete the solution. Of the two possible cyanogen bromide sequences shown above, only the first could have arginine at position 4. The sequence is

CB3–CB4–CB2–CB1 = Ser–Gly–Met–Arg–Thr–Met–Val–Met–Ser–Glu.

Problem 2.7

In this problem two small peptides are joined by a pair of disulfide bonds. We shall reduce the peptide to obtain the individual components, solve each of them separately, and finally deduce the locations of the disulfide bonds.

Reduction of the disulfide bonds in peptide X gave peptides Y and Z.

Y = (Asp,Cys_2,Glu,Gly,His,Ile,Lys_2,Phe,Pro_2,Ser,Thr_2,Trp,Val).

Two of the tryptic fragments have *C*-terminal lysine; thus YT1, which has no basics, must be at the *C*-terminus of Y (see Figure 1).

YT1 = (Asp,His,Thr) = *C*-terminus
YT2 = (Glu,Pro,Thr)–Lys
YT3 = (Cys_2,Gly,Ile,Phe,Pro,Ser,Trp,Val)–Lys

Y = (YT2,YT3)-YT1 =

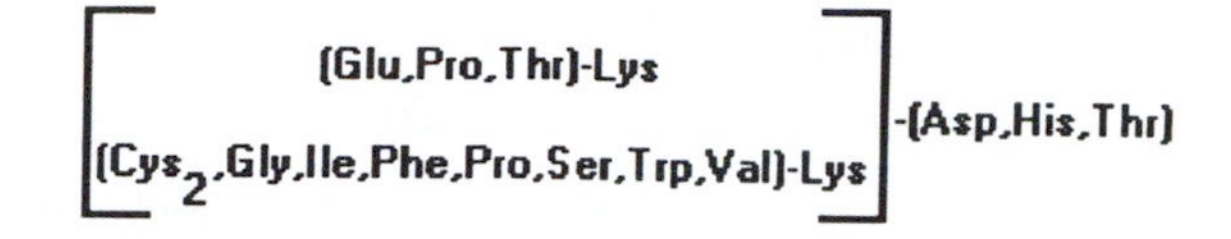

FIGURE 1 Partial tryptic sequence of peptide Y.

YCT3 is clearly the *C*-terminal chymotryptic fragment (Figure 2):

YCT1 = (Ile,Val)–Trp
YCT2 = (Cys,Gly,Pro,Ser)–Phe
YCT3 = (Cys,Glu,Lys$_2$,Pro,Thr)–(Asp,His,Thr)

Y = (YCT1,YCT2)-YCT3 =

FIGURE 2 Partial chymotryptic sequence of peptide Y.

The *C*-terminal fragment consists of residues 9 to 17. The first part of that fragment contains both lysine residues. If the tryptic sequence were YT2–YT3 the lysine in YT2 would be at position 4 of peptide Y. This is inconsistent with what we have just seen. Consequently, the tryptic sequence must be Y = YT3–YT2–YT1 = (Cys$_2$,Gly,Ile,Phe,Pro,Ser,Trp,Val)–Lys–(Glu,Pro,Thr)–Lys–(Asp,His,Thr).

The six amino acid residues at the *N*-terminus of YCT3 are (Cys,Glu,Lys$_2$,Pro,Thr). Since all of these but Cys are accounted for in positions 10 to 14 in the tryptic sequence, it follows that one of the Cys residues must be at position 9, immediately preceding Lys$_{10}$. Thus: Y = YT3–YT2–YT1 =

(Cys,Gly,Ile,Phe,Pro, Ser,Trp,Val)–Cys–Lys–(Glu,Pro,Thr)–Lys–(Asp,His,Thr).

The chymotryptic sequence can be shown as in Figure 3.

Y = (YCT1,YCT2)-YCT3 =

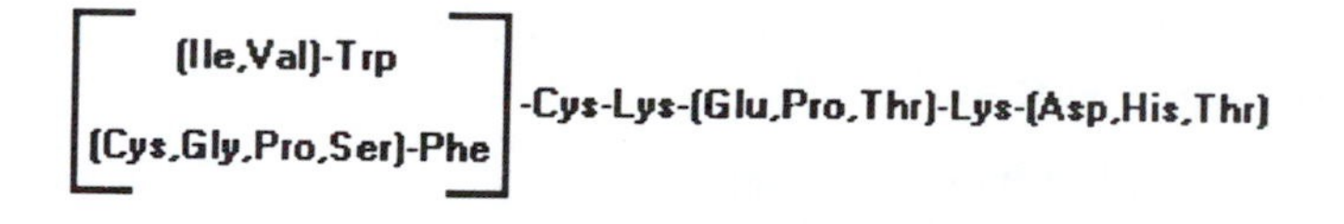

FIGURE 3 Partial chymotryptic sequence of peptide Y.

The elastase fragments are

YE1 = Ser–Gly
YE2 = (Cys,Phe,Pro)–Val
YE3 = (Asp,Cys,Glu,His,Ile,Lys_2,Pro,Thr_2,Trp)

YE3 lacks the target amino acids, hence it is at the *C*-terminus. Since the *C*-terminal nonapeptide is partially known from the chymotryptic sequence shown above, YE3 = (Ile,Trp)–Cys–Lys–(Glu,Pro,Thr)–Lys–(Asp,His,Thr).

The fact that the grouping (Ile,Trp) immediately precedes Cys_9 establishes the chymotryptic sequence as Y = YCT2–YCT1–YCT3 = (Cys,Gly,Pro,Ser)–Phe–(Ile,Val)–Trp–Cys–Lys–(Glu,Pro,Thr)–Lys–(Asp,His,Thr). Since phenylalanine must be at position 5, the sequence of elastase fragments must be Y = YE1–YE2–YE3 = Ser–Gly–(Cys,Pro)–Phe–Val–Ile–Trp–Cys–Lys–(Glu,Pro,Thr)–Lys–(Asp,His,Thr).

One of the peptic fragments appears to be anomalous.

YP1 = Phe–Val–Ile
YP2 = Ser–Gly–(Cys,Pro)
YP3 = Trp–Cys–Lys–Thr
YP4 = (Asp,Glu,His,Lys,Pro,Thr)

From the lithium borohydride data, YP2 = Ser–Gly–Pro–Cys, and Y =

Ser–Gly–Pro–Cys–Phe–Val–Ile–Trp–Cys–Lys–Thr–(Asp,Glu,His,Lys,Pro,Thr)

Fragment YP4 contains both aspartate and glutamate. It is, of course, impossible for both of these to appear in the same peptic fragment; one of them must be present as the amide. Since pepsin cleaved between positions 11 and 12, residue 12 must be one of the target amino acids of pepsin. From the elastase sequence we identify that amino acid as glutamate. Hence the aspartate which appears later must have been present as asparagine: Y = Ser–Gly–Pro–Cys–Phe–Val–Ile–Trp–Cys–Lys–Thr–Glu–Pro–Lys–(Asn,His,Thr).

Treatment of Y with amidase converts asparagine to aspartate and creates a new pepsin cleavage site. As a result, fragment YAP1 = Asp–His, and the entire sequence is

Y = Ser–Gly–Pro–Cys–Phe–Val–Ile–Trp–Cys–Lys–Thr–Glu–Pro–Lys–Thr–Asn–His

Now we can proceed to fragment Z.

Z = (Ala_2,Arg_2,Asp,Cys_2,His,Ile,Met_2,Phe, Pro,Ser_2,Thr,Tyr).

Cyanogen bromide fragment ZCB1, which lacks homoserine, must be at the *C*-terminus (Figure 4):

ZCB1 = (Ala,Thr) = *C*-terminus
ZCB2 = His–Met

Z = (ZCB2,ZCB3)-ZCB1 =

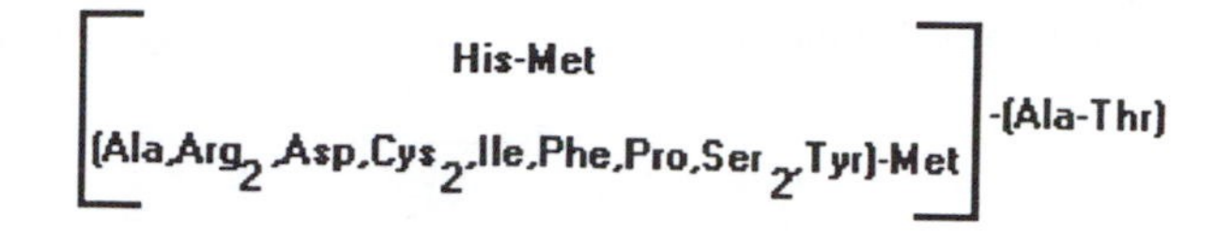

FIGURE 4 Partial cyanogen bromide sequence of peptide Z.

ZCB3 = (Ala,Arg_2,Asp,Cys_2,Ile,Phe,Pro,Ser_2,Tyr)–Met

Chymotryptic fragment ZCT2 is clearly at the *C*-terminus. Since (Ala,Thr) has already been established at the *C*-terminus in the cyanogen bromide sequence, and since methionine must immediately precede (Ala,Thr), the *N*-terminus of ZCT2 must be cysteine.

ZCT1 = (Ala,Ile)–Tyr
ZCT2 = Cys–Met–(Ala,Thr) = *C*-terminus
ZCT3 = (Arg_2,Asp,Cys,His,Met,Pro,Ser_2)–Phe

In the cyanogen bromide fragments, only ZCB3 contains cysteine. The sequence must then be Z = ZCB2–ZCB3–ZCB1 = His–Met–(Ala,Arg_2,Asp,Cys, Ile,Phe,Pro,Ser_2,Tyr)–Cys–Met–(Ala,Thr). The *N*-terminal dipeptide His–Met appears in ZCT3; therefore, this must be the *N*-terminal chymotryptic fragment:

ZCT3 = His–Met–(Arg_2,Asp,Cys,Pro,Ser_2)–Phe and Z = ZCT3–ZCT1–ZCT2 = His–Met–(Arg_2,Asp,Cys,Pro,Ser_2)–Phe–(Ala,Ile)–Tyr–Cys–Met–(Ala,Thr).

The tryptic sequence is readily established. ZT3 lacks basic amino acids and must be at the *C*-terminus, while ZT2 contains the *N*-terminal sequence His–Met:

ZT1 = (Asp,Pro,Ser)–Arg
ZT2 = His–Met–Ser–Arg = *N*-terminus
ZT3 = (Ala_2,Cys_2,Ile,Met,Phe,Thr,Tyr) = *C*-terminus

Thus Z = ZT2–ZT1–ZT3 =

His–Met–Ser–Arg–(Asp,Pro,Ser)–Arg–(Ala_2,Cys_2,Ile,Met,Phe,Thr,Tyr).

At the *C*-terminus, positions 10 to 17 have been shown to be Phe–(Ala,Ile)–Tyr–Cys–Met–(Ala,Thr). The only amino acid remaining in ZT3 is cysteine, which must therefore be at the *N*-terminus of ZT3: ZT3 = Cys–Phe–(Ala,Ile)–Tyr–Cys–Met–(Ala,Thr). Hence Z = ZT2–ZT1–ZT3 = His–Met–Ser–Arg–(Asp,Pro,Ser)–Arg–Cys–Phe–(Ala,Ile)–Tyr–Cys–Met–(Ala,Thr).

The remaining ambiguities are eliminated when we consider the peptic digest:

ZP1 = Phe–(Ala,Ile)

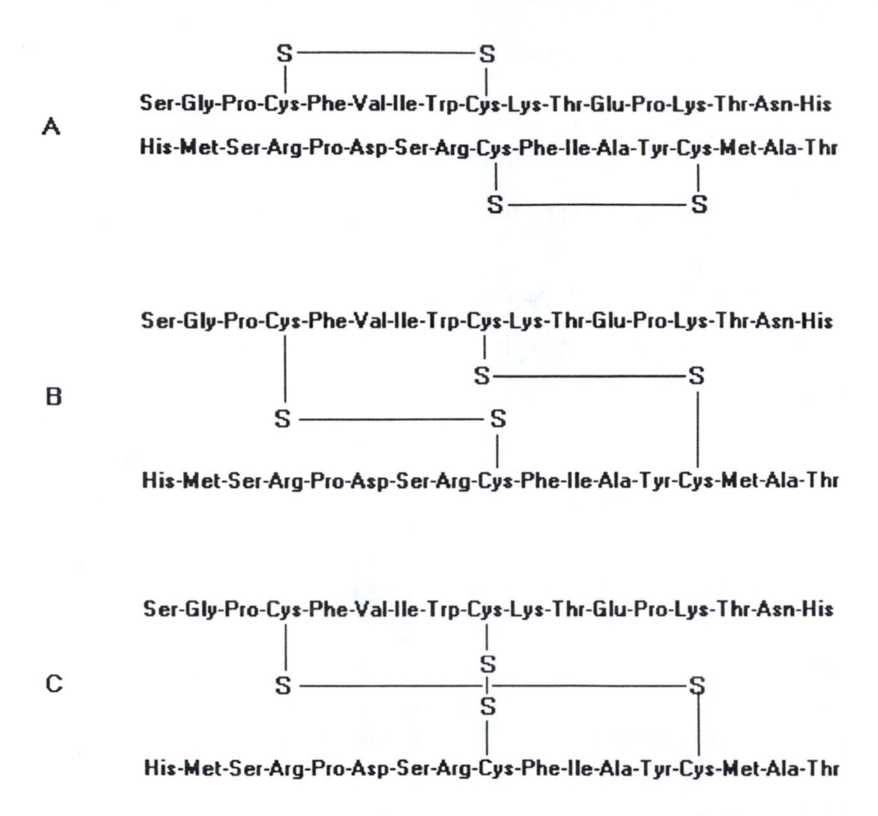

FIGURE 5 Possible disulfide arrangements in peptide X.

ZP2 = Asp–(Arg,Cys,Ser)
ZP3 = Tyr–(Ala,Cys,Met,Thr)
ZP4 = His–Met–Ser–Arg–Pro = *N*-terminus

Fragment ZP4, which lacks the target amino acids of pepsin, is the *N*-terminus. Positions 1 to 4 have already been established. Position 5 must therefore contain proline. Then positions 6 and 7 in the tryptic sequence (Asp,Ser) must be in the sequence Asp–Ser, found in ZP2. Since positions 8 and 9 are known to be Arg–Cys, the sequence of ZP2 is Asp–Ser–Arg–Cys. The *C*-termini of ZP1 (Ala) and ZP3 (Thr) are established by lithium borohydride reduction so that ZP1 = Phe–Ile–Ala, ZP3 = Tyr–Cys–Met–Ala–Thr, and Z = ZP4–ZP2–ZP1–ZP3 =

His–Met–Ser–Arg–Pro–Asp–Ser–Arg–Cys–Phe–Ile–Ala–Tyr–Cys–Met–Ala–Thr

The only remaining problem is the determination of the locations of the disulfide bonds. Since there are four sulfhydryl groups, there are exactly three ways in which two disulfide bonds can form. These are shown in Figure 5. Of these, one arrangement involves a pair of intrachain disulfides (Y4–Y9/Z9–Z14, i.e., one disulfide bond joins residues 4 and 9 in fragment Y, while the other joins residues 9 and 14 in fragment Z). This is shown in model A.

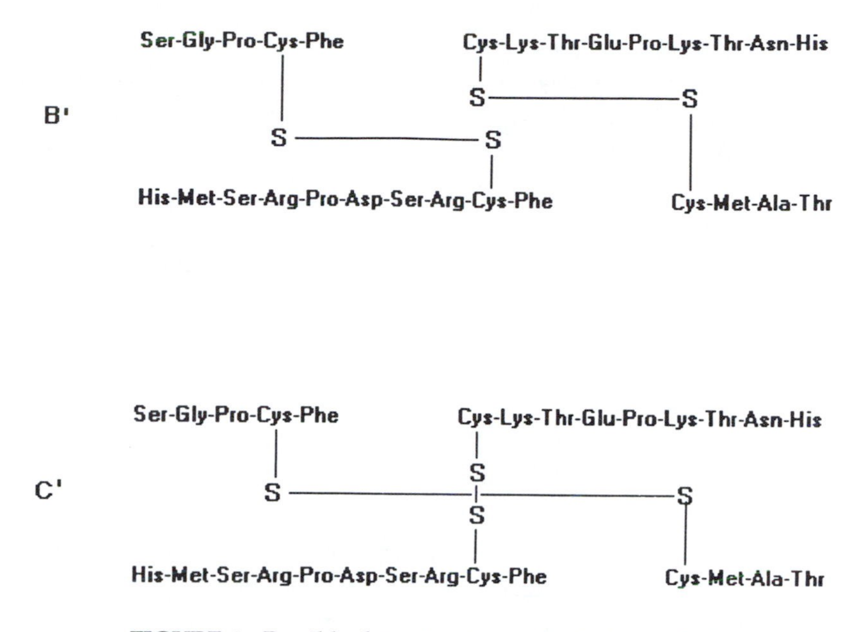

FIGURE 6 Possible chymotryptic products of peptide X.

Although there is nothing theoretically wrong with this structure, it should be apparent that this really represents two separate polypeptides and that there is no reason to expect them to behave as a single molecule as required by the data. Thus model A can be discarded.

The structure of peptide X, then, must contain two interchain disulfide bonds. The two possibilities are shown in models B (Y4–Z9/Y9–Z14) and *C* (Y4–Z14/Y9–Z9).

Chymotryptic digestion of X removes residues 6 to 8 from fragment Y and residues 11 to 13 from fragment Z, leaving two products, each of which contains a disulfide bond. These are depicted in models B′ and C′ in Figure 6.

We are told that each of the four products resulting from the chymotrypsin treatment contains an aromatic amino acid, but of the fragments shown in model B′, one contains two aromatic amino acids while the other contains none. Thus only model C′ (hence *C*) is consistent with the data, and in peptide X the disulfide arrangement is Cys_4(Y)–Cys_{14}(Z) and Cys_9(Y)–Cys_9(Z).

Problem 2.8

Simultaneous reduction of all the peptic fragments with lithium borohydride yields only 2-aminoethanol, indicating that all seven fragments had *C*-terminal glycine. Combining this with our understanding of the specificity of pepsin, we can show the following partial sequences:

P1 = (Ala,Ser)–Gly

P2 = Phe–(Ala,Ser)–Gly
P3 = Trp–(Ala,Ser)–Gly
P4 = Tyr–(Ala,Ser)–Gly
P5 = Asp–(Lys,Ser)–Gly
P6 = Leu–(Lys,Ser)–Gly
P7 = Glu–(Arg,Ser_2)–Gly

P1 lacks the target amino acids, so we place it at the *N*-terminus of peptide E.

From the FDNB experiment we conclude that all four chymotryptic fragments have *N*-terminal alanine:

CT1 = Ala–(Gly,Ser)
CT2 = Ala–(Asp,Gly_2,Lys,Ser_2)–Tyr
CT3 = Ala–(Gly_2,Leu,Lys,Ser_2)–Trp
CT4 = Ala–(Arg,Glu,Gly_2,Ser_3)–Phe

CT1 lacks aromatic amino acids, hence it must be at the *C*-terminus of the original peptide. Since we already know, from the peptic digest, that the *C*-terminal amino acid is glycine, we can show the sequence CT1 = Ala–Ser–Gly = *C*-terminal. Further, since the chymotryptic digest shows the *N*-terminal amino acid to be alanine, we can show the sequence P1 = Ala–Ser–Gly = *N*-terminal. We can show a partial chymotryptic sequence as in Figure 7.

(CT2,CT3,CT4)-CT1 =

[Ala-(Asp,Gly_2,Lys,Ser_2)-Tyr | Ala-(Gly_2,Leu,Lys,Ser_2)-Trp | Ala-(Arg,Glu,Gly_2,Ser_3)-Phe] -Ala-Ser-Gly

FIGURE 7 Partial chymotryptic sequence of peptide E.

The results of the tryptic digest can be shown as:

T1 = (Ala,Gly_2,Ser,Trp)
T2 = (Ala,Glu,Gly,Ser_2)–Arg
T3 = (Ala,Gly_2,Leu,Ser_2,Tyr)–Lys
T4 = (Ala,Asp,Gly_2,Phe,Ser_3)–Lys

Fragment T1 lacks basic amino acids, therefore, it must be at the *C*-terminus. We already know the identities of the amino acids at positions 26 to 28, so that we can show T1 as (Gly,Trp)–Ala–Ser–Gly (positions 24 to 28).

The grouping (Gly,Trp) at positions 24 and 25 must come from CT3 (the only chymotryptic fragment containing tryptophan): CT3 = Ala–(Gly,Leu,Lys,Ser_2)–Gly–Trp. Since CT1 is already known to be at the *C*-terminus, CT3–CT1 = Ala–(Gly,Leu,Lys,Ser_2)–Gly–Trp–Ala–Ser–Gly (positions 18 to 28).

Only one tryptic peptide, T3, contains leucine; T3 must therefore precede T1. T3 = (Ala,Gly_2,Leu,Ser_2,Tyr)–Lys and T3–T1 = (Ala,Gly_2,Leu,Ser_2,Tyr)–Lys–(Gly,Trp)–Ala–Ser–Gly. Combining what we have seen for positions 18 to 28 in the chymotryptic sequence with the tryptic data, we can show that T3–T1 = (Gly,Tyr)–Ala–(Gly,Leu,Ser_2)–Lys–Gly–Trp–Ala–Ser–Gly (positions 16 to 28). The only chymotryptic fragment containing tyrosine is CT2, which must come right before CT3. CT2 = Ala–(Asp,Gly_2,Lys,Ser_2)–Tyr and CT2–CT3–CT1 = Ala–(Asp,Gly,Lys,Ser_2)–Gly–Tyr–Ala–(Gly,Leu,Ser_2)–Lys–Gly–Trp–Ala–Ser–Gly (positions 10 to 28).

Aspartate is found in tryptic fragment T4. T4 = (Ala,Asp,Gly_2,Phe,Ser_3)–Lys = (Gly,Phe,Ser)–Ala–(Asp,Gly,Ser_2)–Lys. Inserting T4 just prior to T3 gives, for positions 7 to 28: T4–T3–T1 = (Gly,Phe,Ser)–Ala–(Asp,Gly,Ser_2)–Lys–Gly–Tyr–Ala–(Gly,Leu,Ser_2)–Lys–Gly–Trp–Ala–Ser–Gly.

The only remaining chymotryptic fragment is CT4 = Ala–(Arg,Glu,Gly_2,Ser_3)–Phe = Ala–(Arg,Glu,Gly,Ser_2)–(Gly,Ser)–Phe. CT4 must be the *N*-terminal chymotryptic fragment, so that CT4–CT2–CT3–CT1 (positions 1 to 28) = Ala–Ser–Gly–(Arg,Glu,Ser)–(Gly,Ser)–Phe–Ala–(Asp,Gly,Ser_2)–Lys–Gly–Tyr–Ala–(Gly,Leu,Ser_2)–Lys–Gly–Trp–Ala–Ser–Gly. Note that positions 1 to 3 were established by the peptic cleavage data, which showed that

P1 = Ala–Ser–Gly.

Inserting the remaining tryptic fragment, T2 = (Ala,Glu,Gly,Ser_2)–Arg = Ala–Ser–Gly–(Glu,Ser)–Arg, at the *N*-terminus gives CT4–CT2–CT3–CT1 = T2–T4–T3–T1 = Ala–Ser–Gly–(Glu,Ser)–Arg–(Gly,Ser)–Phe–Ala–(Asp,Gly,Ser_2)–Lys–Gly–Tyr–Ala–(Gly,Leu,Ser_2)–Lys–Gly–Trp–Ala–Ser–Gly.

The remainder of the sequence can be determined from the peptic fragments:

P1 = Ala–Ser–Gly (positions 1 to 3)
P7 = Glu–Ser–Arg–Ser–Gly (positions 4 to 8)
P2 = Phe–Ala–Ser–Gly (positions 9 to 12)
P5 = Asp–Ser–Lys–Gly (positions 13 to 16)
P4 = Tyr–Ala–Ser–Gly (positions 17 to 20)
P6 = Leu–Ser–Lys–Gly (positions 21 to 24)
P3 = Trp–Ala–Ser–Gly (positions 25 to 28)

The entire sequence is CT4–CT2–CT3–CT1 = T2–T4–T3–T1 = P1–P7–P2–P5–P4–P6–P3 = Ala–Ser–Gly–Glu–Ser–Arg–Ser–Gly–Phe–Ala–Ser–Gly–Asp–Ser–Lys–Gly–Tyr–Ala–Ser–Gly–Leu–Ser–Lys–Gly–Trp–Ala–Ser–Gly.

Problem 2.9

Although the tone of this problem is playful, its purpose is quite serious. Here, for the first time, is an open-ended question, one which has no single correct answer. The thing that should be immediately obvious is that the pentapeptide must contain at least two different amino acids, one of which is a target for the enzyme chosen. Among the possible sequences that can be established by a single cleavage are:

Lys–Ala–Ala–Ala–Ala (cleavage by trypsin)
Ala–Arg–Ser–Ser–Ser (cleavage by trypsin)
Gly–Gly–Phe–Gly–Gly (cleavage by chymotrypsin)
Val–Val–Val–Tyr–Gly (cleavage by chymotrypsin)
Ile–Ile–Trp–Ser–Ser (cleavage by chymotrypsin)
Ala–Glu–Val–Val–Val (cleavage by pepsin)
Met–Met–Leu–Asn–Asn (cleavage by pepsin)

Problem 2.10

This problem is also unusual in that it provides the opportunity to work backward, deriving the data from the known amino acid sequence. The idea is to give a feeling for how sequencing problems are written.

The data, of course, are fairly straightforward, and may be presented in any order:

Tryptic fragments	T1 = Gly
	T2 = (Arg,Pro)
	T3 = (Asp,Cys,Glu,Lys,Phe,Tyr)
Peptic fragments	P1 = Cys
	P2 = Tyr
	P3 = (Glu,Phe)
	P4 = (Arg,Asp,Gly,Lys,Pro)
Chymotryptic fragments	CT1 = Phe
	CT2 = (Cys,Tyr)
	CT3 = (Arg,Asp,Glu,Gly,Lys,Pro)

Fragment T1 (Gly) is clearly the *C*-terminus, so a partial tryptic sequence can be shown as in Figure 8.

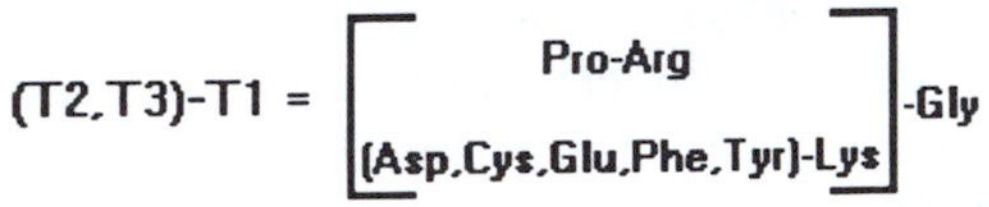

(T2,T3)-T1 = [Pro-Arg / (Asp,Cys,Glu,Phe,Tyr)-Lys]-Gly

FIGURE 8 Partial tryptic sequence of the witch doctor's peptide.

Peptic fragment P1 (Cys) is at the *N*-terminus; consequently, we can place T3 at the *N*-terminus: T3 = Cys–(Asp,Glu,Phe,Tyr)–Lys, and the tryptic sequence can be shown as T3–T2–T1 = Cys–(Asp,Glu,Phe,Tyr)–Lys–Pro–Arg–Gly. The Glu in fragment P3 must be glutamine: P3 = Phe–Gln. Fragment P4, which contains glycine, is at the *C*-terminus, and the partial peptic sequence is shown in Figure 9.

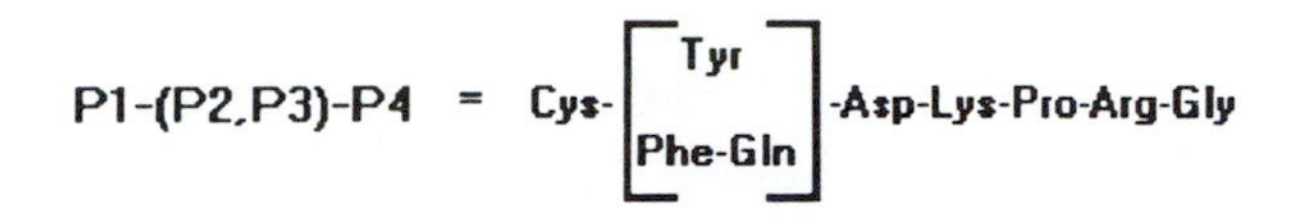

FIGURE 9 Partial peptic sequence of the witch doctor's peptide.

Chymotryptic fragment CT2 (Cys–Tyr) is at the *N*-terminus; this establishes the entire sequence: T3–T2–T1 = P1–P2–P3–P4 = CT2–CT1–CT3 =

Cys–Tyr–Phe–Gln–Asp–Lys–Pro–Arg–Gly

Chapter 3

Problem 3.1

A. GTC TGA TCG TAG CTG
B. TCG GTC GAT CGT GAC TGA TCG TGT A
C. TGC TGC CGA TCG TAG TCA GTC ACG TAC GTG CAT CC

Problem 3.2

A. TCA GCT ACG ATT CAG
B. GTA GCT ACG CAA CTT ACG GGG ACT G
C. CGA TCG ACG TTA GCG GCG ACT TTA GCA CAG CAA GG

Chapter 4

Problem 4.1

There is a single restriction site for each of the enzymes. The only question to be resolved is the relative location of each of the sites. We can start by assigning the site of either of the enzymes. Let us arbitrarily choose Ava I. Its restriction site is 1.0 kbp from one end of the molecule and 1.8 kbp from the other. Let us arbitrarily select 1.0 kbp as the Ava I cleavage site.

Bgl II, which yields fragments of 0.3 and 2.7 kbp, cleaves at either 0.3 or 2.7 kbp. In the double digest, the 1.0 kbp Ava I fragment is cleaved to fragments of 0.3 and 0.7 kbp; thus the Bgl II site is within the 1.0 kbp Ava I fragment. If Bgl II cleaves within the 1.0 kbp Ava I fragment, its cleavage site must be at 0.3 kbp. Thus the entire map shows one restriction site for Ava I at 1.0 kbp, and one site for Bgl II at 0.3 kbp. Had we selected 1.8 kbp as the location of the Ava I site, then the Bgl II site would have turned out to be at 2.5 kbp. These two maps are shown in Figure 10.

You should be able to convince yourself that these two maps are really a pair of mirror images, and therefore identical. It may help to note that on the left hand map the order of fragments is 0.3, 0.7, 1.8 kbp from "left" to "right"; while on the right hand map the order of fragments is 1.8, 0.7, 0.3 kbp, from "left" to "right", or 0.3, 0.7, 1.8 kbp from "right" to "left". Since the only

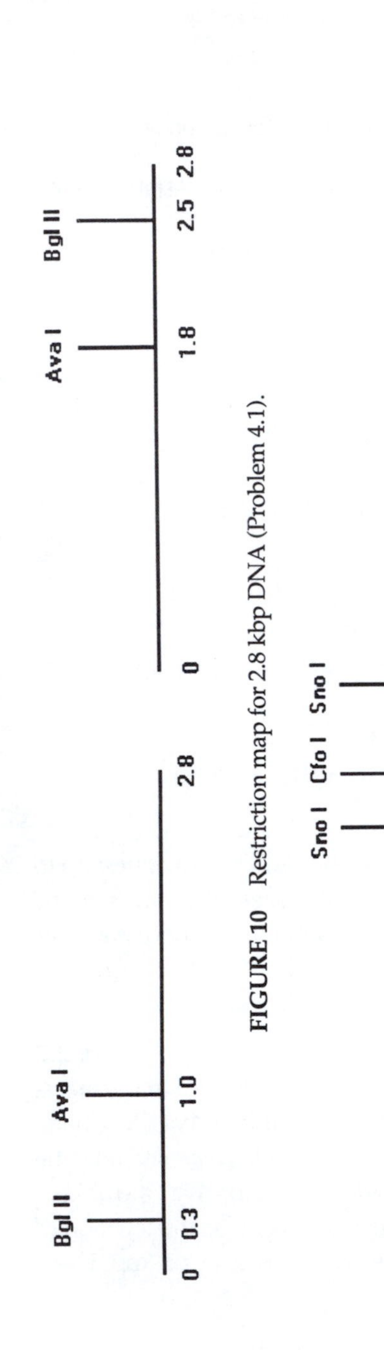

FIGURE 10 Restriction map for 2.8 kbp DNA (Problem 4.1).

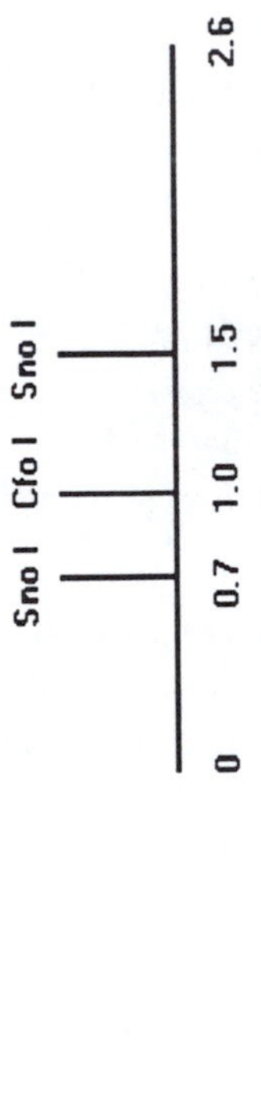

Sno I Cfo I Sno I

0 0.7 1.0 1.5 2.6

FIGURE 11 Restriction map for 2.6 kbp DNA (Problem 4.2).

difference between the two is the direction in which the fragments are read, they are mirror images.

Problem 4.2

Assume the Cfo I site to be at 1.0 kbp. The double digest data show that the Cfo I site is within the 0.8 kbp Sno I fragment and that each of the two Cfo I fragments is cleaved by Sno I (the 1.0 kbp fragment yields fragments of 0.3 and 0.7 kbp, while the 1.6 kbp fragment is cleaved to fragments of 0.5 and 1.1 kbp).

The Sno I site in the 1.0 kbp Cfo I fragment is at either 0.3 or 0.7 kbp. But cleavage at 0.3 kbp would yield a 0.3 kbp fragment in the Sno I single digest. Since this is not found, the first Sno I site must be at 0.7 kbp.

The second Sno I site can be either at 1.5 kbp or at 2.1 kbp. Cleavage at 2.1 kbp would give a fragment of 0.5 kbp in the Sno I single digest. This is not found; hence the second Sno I site is at 1.5 kbp.

The entire map, then, shows Sno I sites at 0.7 and 1.5 kbp, and a Cfo I site at 1.0 kbp, as shown in Figure 11.

Problem 4.3

The 0.7 kbp Pst I fragment in the single digest is replaced by fragments of 0.3 and 0.4 kbp in the double digest (note that the Pst I fragments of 0.5, 0.9, and 1.4 kbp appear in both the single and double digests). Thus the Hind III site is within the 0.7 kbp Pst I fragment.

The 0.8 kbp Hind III fragment is further cleaved by Pst I to give fragments of 0.3 and 0.5 kbp, while the 2.7 kbp Hind III fragment is broken down to fragments of 0.4, 0.9, and 1.4 kbp. This implies two Pst I sites within the 2.7 kbp Hind III fragment.

Assume the Hind III site to be at 0.8 kbp. The first Pst I site must be within the 0.8 kbp Hind III fragment, such that the double digest will produce fragments of 0.3 and 0.5 kbp. It should be clear that the Pst I site cannot be at 0.3 kbp, since the Pst I single digest shows no 0.3 kbp fragment. The first Pst I site, therefore, must be at 0.5 kbp.

The second Pst I site must be either 0.4, 0.9, or 1.2 kbp from the Hind III site, i.e., at either 1.2, 1.7, or 2.2 kbp. But we have established that the Hind III site must be within a 0.7 kbp Pst I fragment. The only way this can occur is for the second Pst I site to be at 1.2 kbp.

The remaining Pst I site is undefined; it may be at either 2.1 or 2.6 kbp. In either case, fragments of 0.9 and 1.4 kbp would appear in the double digest.

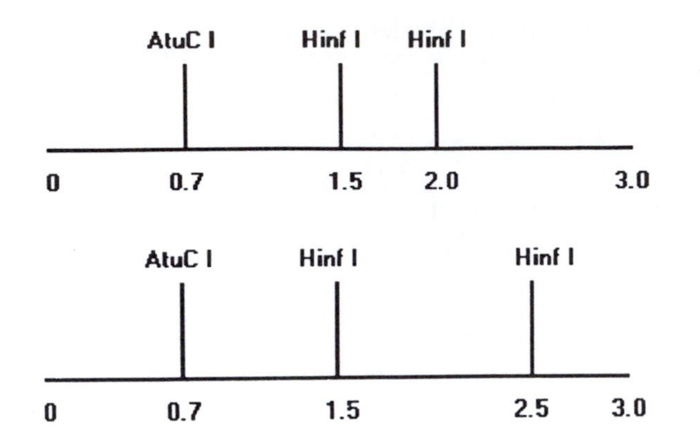

FIGURE 12 Possible restriction maps for 3.0 kbp DNA (Problem 4.4).

Problem 4.4

The AtuC I site is within the 1.5 kbp Hinf I fragment, whereas both Hinf I sites are within the 2.3 kbp AtuC I fragment. If we assume that AtuC I cleaves at 0.7 kbp, then the first Hinf I site must be at 1.5 kbp. This is the only position which satisfies both requirements noted above. The remaining Hinf I site is at either 2.0 or 2.5 kbp, so there are two possible maps to describe the data (Figure 12).

Gsb I has a single restriction site at either 0.3 or 2.7 kbp, which must be within the 0.5 kbp Hinf I fragment. If the second Hinf I site is at 2.0 kbp, the Gsb I cleavage at either 0.3 or 2.7 kbp would be outside the 0.5 kbp Hinf I fragment. Consequently, the second Hinf I site must be at 2.5 kbp. The entire map is shown in Figure 13.

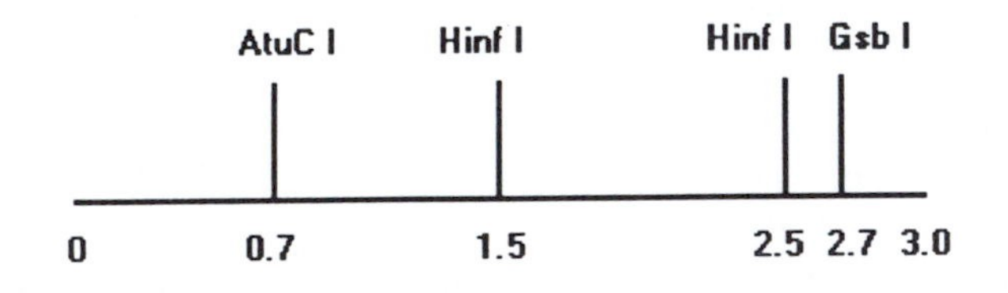

FIGURE 13 Restriction map for 3.0 kbp DNA (Problem 4.4).

Problem 4.5

Start by assigning the location of the BamH I site (any site can be chosen first). Since BamH I yields fragments of 1.2 and 3.8 kbp, locate the restriction site at 1.2 kbp from the left end of the double stranded fragment (as usual, it would make no difference if the site were placed at 1.2 kbp from the right end, since the fragment is double stranded). All other restriction sites will be located relative to the BamH I site.

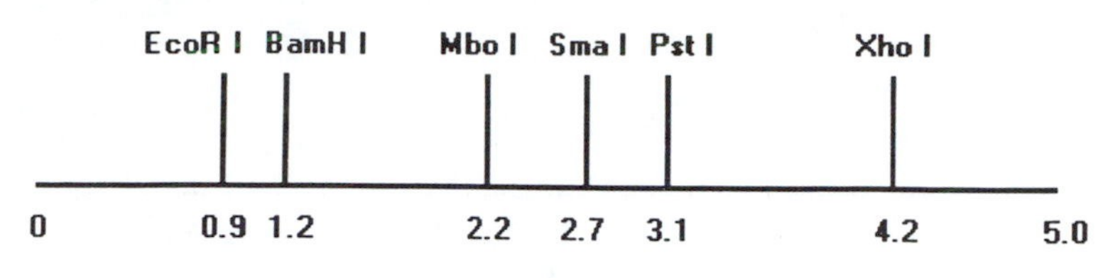

FIGURE 14 Restriction map for 5.0 kbp DNA (Problem 4.5).

The EcoR I site may be at 0.9 or 4.1 kbp, but it must be within the 1.2 kbp BamH I fragment. Thus EcoR I cleaves at 0.9 kbp. The Mbo I site at 2.2 or 2.8 kbp must be within the 3.8 kbp BamH I fragment. Conversely, the BamH I site at 1.2 kbp must be within the 2.2 kbp Mbo I fragment. Hence, Mbo I must cleave at 2.2 kbp.

Sma I cleaves at 2.3 or 2.7 kbp, within the 4.1 kbp EcoR I fragment. Since the EcoR I site (known to be at 0.9 kbp) must be within the 2.7 kbp Sma I fragment, it is clear that Sma I must cleave at 2.7 kbp. The Pst I site at 1.9 or 3.1 kbp is within the 2.3 kbp Sma I fragment. Since the 2.3 kbp Sma I fragment extends from 2.7 to 5.0 kbp, the Pst I site must be within this region, i.e., at 3.1 kbp.

The Xho I site is at 0.8 or 4.2 kbp, within the 2.8 kbp Mbo I fragment. Only cleavage at 4.2 kbp satisfies this requirement. The map is shown in Figure 14.

Each of the multiple digests will yield two fragments, one from each end of the double stranded DNA:

Enzyme	Fragment 1	Fragment 2
EcoR I/Xho I	0.8	0.9
BamH I/Pst I	1.2	1.9
BamH I/Sma I/Pst I	1.2	1.9

Problem 4.6

The EcoR I and Hpa I data indicate a single restriction site for each enzyme, yielding a linear rather than a circular molecule. The fact that each enzyme cleaves at all is confirmed by the double digest data. If, for example, EcoR I did not cleave the original molecule, then it would not act on any of the Alu I fragments. In fact, linear and circular DNAs can be distinguished by their electrophoretic behaviors so that there is no doubt that both EcoR I and Hpa I act on the molecule. Alu I, which yields three fragments, must have three restriction sites. Note that this differs from the situation in a linear DNA in which three fragments would imply only two restriction sites.

Since the DNA is circular, one must arbitrarily select some point in the molecule to call the "beginning" or "origin." Let us choose the EcoR I site and calculate all cleavages with reference to this position. Locations on the map will be given in kilobase pair distances from the origin in a clockwise direction, so we can say that the EcoR I site is at position 0.

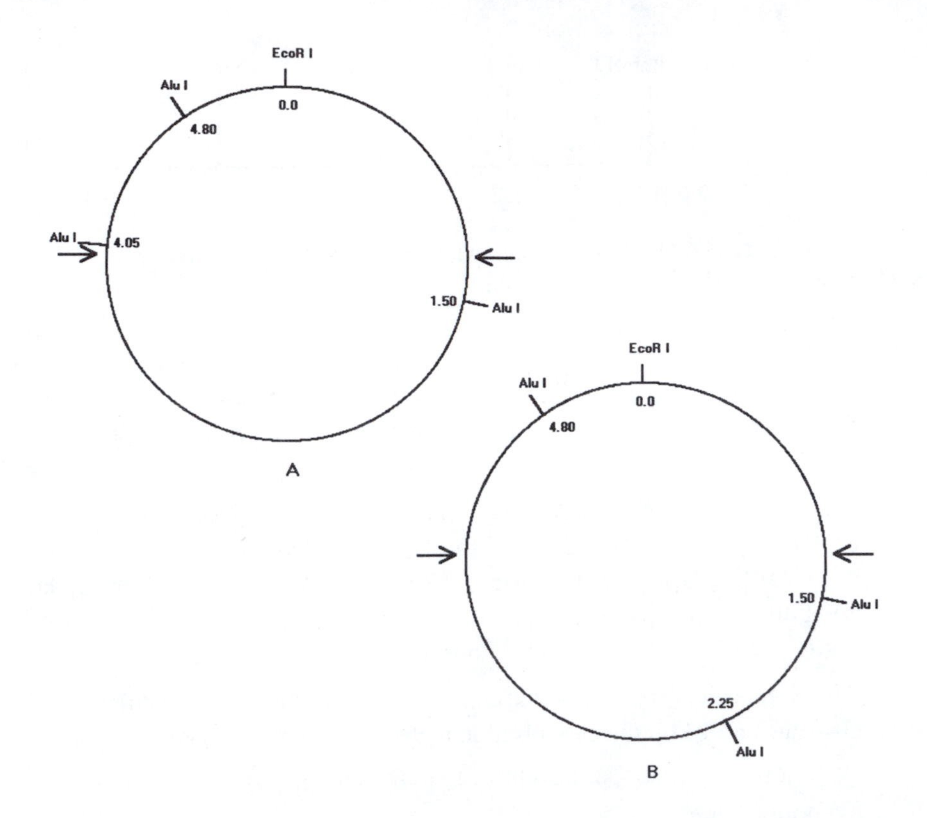

FIGURE 15 Possible restriction maps for 5.3 kbp circular DNA (Problem 4.6).

Data from the Alu I/EcoR I double digest indicate that the EcoR I site is within the 2.00 kbp Alu I fragment. There are two possible orientations for the Alu I sites relative to the EcoR I site, as shown in Figure 15.

In both maps A and B, the EcoR I site is flanked by two Alu I sites (at 1.50 and 4.80 kbp), so that the double digest yields fragments of 0.50 and 1.50 kbp. Note that it is irrelevant whether the 0.50 kbp fragment is shown "to the left" of the EcoR I site or "to the right," since flipping the map would yield the opposite orientation.

Two of the Alu I sites have thus been accounted for; the third remains ambiguous. In map A, Alu I sites are shown at 1.50, 4.05, and 4.80 kbp, so that the 0.75 kbp fragment (positions 4.05 to 4.80) is between the 2.55 and 0.50 kbp fragments. In map B, Alu I sites are at 1.50, 2.25, and 4.80 kbp, placing the 0.75 kbp fragment (positions 1.50 to 2.25) between the 1.50 and 2.55 kbp fragments. The data presented so far do not allow a decision between these two maps.

The problem is resolved by the Hpa I data. An EcoR I/Hpa I double digest yields fragments of 1.30 and 4.00 kbp. The Hpa I site must therefore be 1.30 kbp away from the EcoR I site, i.e., either at position 1.30 or at position 4.00 (arrows in both maps).

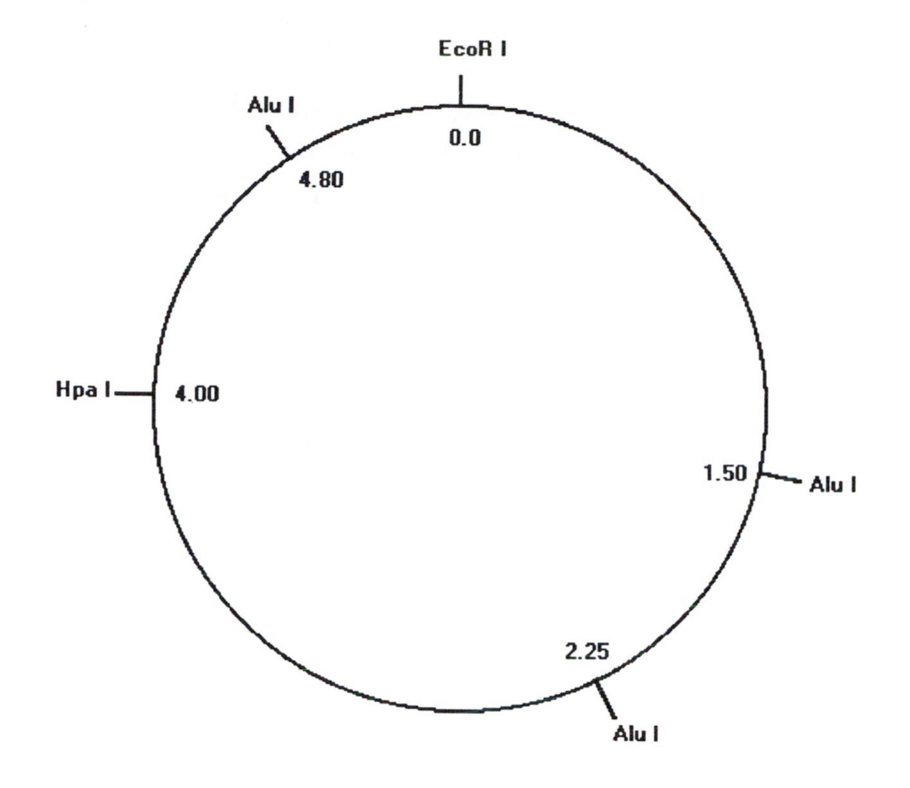

FIGURE 16 Restriction map of 5.3 kbp circular DNA (Problem 4.6).

Digestion of the 1.30 kbp fragment with Alu I gives fragments of 0.80 and 0.50 kbp. In map A, treatment of the 1.30 kbp fragment with Alu I would yield either no new products, if the Hpa I site were at position 1.30, or three products (0.05, 0.75, and 0.50 kbp), if the Hpa I site were at position 4.00. Thus map A cannot describe the DNA in question.

In map B, an Alu I digest of the 1.30 kbp fragment would yield no new products if the Hpa I site were at position 1.30. It is only when the Hpa I site is at position 4.00 that subsequent cleavage with Alu I will yield the observed fragments. Therefore, the entire map must be as shown in Figure 16.

Problem 4.7

A plot of migratory distance as a function of the logarithm of the number of base pairs for the 1 Kb Ladder standards yields a slope of –33.3 and a intercept of 124. The actual measured values are shown in Table 1. The measured migratory distances and calculated sizes of the individual fragments are listed in Table 2.

Problem 4.8

The slope and intercept calculated from the logarithmic plot are –31.4 and 124, respectively. The measured migratory distances and calculated sizes of

Table 1

Electrophoretic Migratory Distances of 1 Kb Ladder DNA Fragments

bp	log (bp)	Distance (mm)
75	1.875	64.8
134	2.127	56.6
154	2.188	55.1
201	2.303	51.1
220	2.342	50.0
298	2.474	45.9
344	2.537	44.1
396	2.598	42.0
517	2.713	38.9
1018	3.008	29.5
1636	3.214	22.8
2036	3.309	19.7
3054	3.485	14.2
4072	3.610	10.5
5090	3.707	7.1
6018	3.779	5.0
7126	3.853	3.0

Table 2

Calculated Sizes of Restriction Fragments Pst I/Hind III Double Digest[a]

Lane 1: Pst I		Lane 2: Hind III		Lane 3: Pst I/Hind III	
Distance	kb	Distance	kb	Distance	kb
25.0	1.4	15.8	2.7	25.0	1.4
31.2	0.9	32.9	0.8	31.1	0.9
34.3	0.7			38.8	0.5
38.8	0.5			42.0	0.4
				45.9	0.3
Sum	3.5		3.5		3.5

[a] Slope = –33.3, intercept = 124.

the individual fragments are listed in Table 3.

Problem 4.9

From the data for the 1 Kb Ladder (well 4) the slope is –31.4 and the intercept is 124. The measured migratory distances and calculated sizes of the individual fragments are listed in Table 4.

The original DNA sample was 4.2 kbp long. Nar I gives fragments of 0.9, 1.2, and 2.1 kbp; Sac I gives fragments of 1.4 and 2.8 kbp; and the double digest shows fragments of 0.5, 0.7, 0.9, and 2.1 kbp. The problem then reduces to the relatively simple case of two restriction sites for Nar I and one site for Sac I. The actual map is shown in Figure 17.

Table 3
Calculated Sizes of Restriction Fragments Hinf I/AtuC I/Gsb I Digest[a]

Lane 1: Hinf I		Lane 2: AtuC I		Lane 3: Gsb I	
Distance	**kb**	**Distance**	**kb**	**Distance**	**kb**
24.1	1.5	18.0	2.3	15.9	2.7
29.5	1.0	34.5	0.7	46.1	0.3
39.0	0.5				
Sum	3.0		3.0		3.0

Lane 4: Hinf I/AtuC I		Lane 5: Hinf I/Gsb I	
Distance	**kb**	**Distance**	**kb**
29.5	1.0	24.2	1.5
33.0	0.8	29.5	1.0
34.7	0.7	46.0	0.3
39.0	0.5	51.3	0.2
Sum	3.0		3.0

[a] Slope = –31.4, intercept = 124.

Table 4
Calculated Sizes of Restriction Fragments Nar I/Sac I Double Digest[a]

Lane 1 Nar I		Lane 2 Sac I		Lane 3 Nar I/Sac I	
Distance	**kb**	**Distance**	**kb**	**Distance**	**kb**
19.4	2.1	15.3	2.8	19.3	2.1
27.1	1.2	25.0	1.4	31.0	0.9
30.7	0.9			34.1	0.7
				38.6	0.5
Sum	4.2		4.2		4.2

[a] Slope = –31.4, intercept = 124.

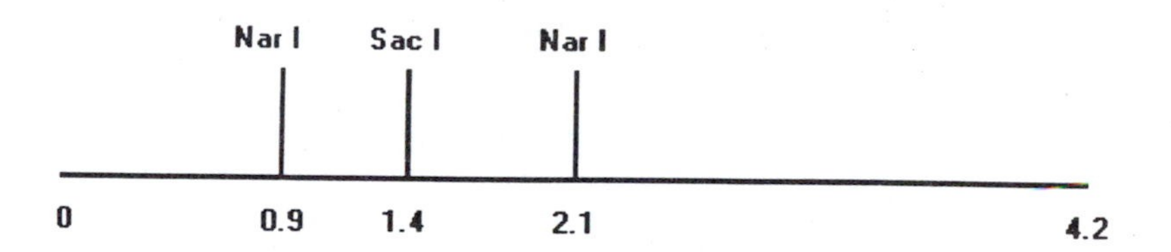

FIGURE 17 Restriction map for 4.2 kbp DNA (Problem 4.9).

Problem 4.10

The standards in this experiment consist of the pGEM DNA Markers. From the data for well 4 the slope is –33.3 and the intercept is 117. The measured

Table 5
Calculated Sizes of Restriction Fragments Mal I/Spc I Double Digest[a]

Lane 1: Mla I		Lane 2: Spc I		Lane 3: Mla I/Spc I	
Distance	**kb**	**Distance**	**kb**	**Distance**	**kb**
3.0	2.6	7.9	1.8	15.1	1.1
21.5	0.7	11.8	1.4	16.9	1.0
29.8	0.4	26.3	0.5	21.5	0.7
				26.4	0.5
				29.8	0.4
Sum	3.7		3.7		3.7

[a] Slope = –33.3, intercept = 117.

migratory distances and calculated sizes of the individual fragments are listed in Table 5.

The double digest data show that the 0.5 kbp Spc I fragment is not cleaved by Mla I, but that the 1.4 kbp Spc I fragment is cleaved to 0.4 and 1.0 kbp, while the 1.8 kbp fragment is cleaved to 0.7 and 1.1 kbp.

The Mla I fragments of 0.4 and 0.7 kbp are not cleaved by Spc I, but the 2.6 kbp fragment is cleaved at two locations, giving fragments of 0.5, 1.0, and 1.1 kbp. Thus we conclude that both Spc I sites are in the 2.6 kbp Mla I fragment.

The 2.6 kbp Mla I fragment does not extend to either end of the DNA molecule. This can be demonstrated in the following way. Suppose that it did include one of the ends of the original 3.7 kbp DNA molecule. For the sake of illustration let us assume that the Mla I cleavage site is 2.6 kbp from the "left" end. The two Spc I sites would have to be "to the left" of position 2.6, while "to the right" would be a 1.1 kbp–long chain.

Recall that Spc I cleaves the 2.6 kbp Mla I fragment to give fragments of 0.5, 1.0, and 1.1 kbp, but that Spc I alone cleaves the DNA to give fragments of 0.5, 1.4, and 1.8 kbp. If a Spc I site were 0.5 kbp "to the left" of the Mla I site, then Spc I cleavage would yield a fragment of 0.5 + 1.1 = 1.6 kbp.

If a Spc I site were 1.0 kbp "to the left" of the Mla I site, then Spc I cleavage would yield a fragment of 1.0 + 1.1 = 2.1 kbp.

If a Spc I site were 1.1 kbp "to the left" of the Mla I site, then Spc I cleavage would yield a fragment of 1.1 + 1.1 = 2.2 kbp.

None of these fragments is found in the Spc I digest. Therefore, the 2.6 kbp Mla I fragment cannot include either end of the DNA molecule. Hence, the 0.4 and 0.7 kbp Mla I fragments must include the ends of the molecule so that one Mla I cleavage site is at 0.4 kbp and the other is at 3.0 kbp. (Placing the sites at 0.7 and 3.3 kbp would yield the mirror image map.)

Recognizing that Spc I has two cleavage sites, yielding three fragments in the single digest, we note that the 0.5 kbp Spc I fragment is not cleaved by Mla I (i.e., the double digest contains a 0.5 kbp fragment). This must mean that one of the Spc I sites is 1.4 kbp from one end of the DNA, the second is 1.8 kbp from the other end, and the two Spc I sites are separated by 0.5 kbp. Since the 1.4 kbp Spc I fragment is cleaved by Mla I to give fragments of 0.4 and 1.0 kbp, the first Spc I site must be at 1.4 kbp. The 1.8 kbp fragment is cleaved to 0.7 and 1.1 kbp, placing the second Spc I site at position 1.9, a distance of 0.5 kbp from the first Spc I site and 1.1 kbp from the Mla I site at 3.0 kbp.

The entire map is shown in Figure 18.

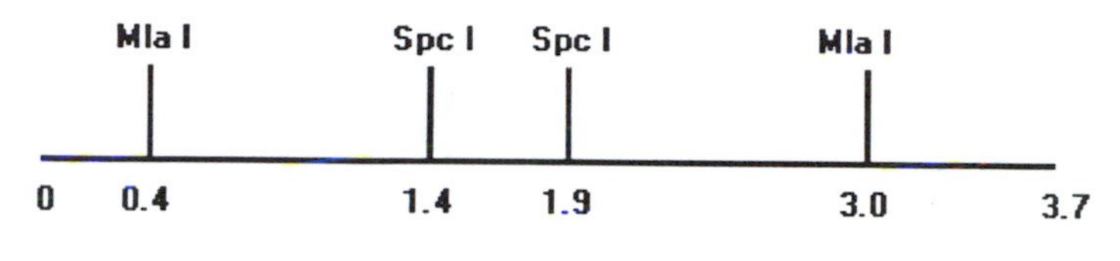

FIGURE 18 Restriction map for 3.7 kbp DNA (Problem 4.10).

Problem 4.11

Data summarizing the mapping experiment are contained in Table 6. Digestion by EcoR I, Bal I, or Ase I, gives only a single band of 4.4 kbp. Thus, either none of these enzymes cleaves the DNA, or the DNA is circular. Since the double digests using these enzymes show cleavage, the first possibility is incorrect, and we must conclude that the DNA is a circular molecule since a single cleavage site on a linear DNA molecule would yield two fragments. (The remaining possibility, namely that all three enzymes cleave at the same point, exactly in the middle of a linear DNA, is also invalidated by the double digest data.)

Two arbitrary assignments must be made. Let us take the EcoR I site as position 0, and let us assume that the Bal I site is 1.4 kbp away from the EcoR I site in the clockwise direction, i.e., the Bal I site is at position 1.4. As we have seen before, we could have assigned position 0 to any site and/or selected the counterclockwise direction; neither choice would have changed the relative positions of subsequent assignments.

Ase I cleaves at either 0.9 or 3.5 kbp; if the Ase I site were at 0.9 kbp, then the Bal I/Ase I double digest would show a fragment of 0.5 kbp. Since this is not found, the Ase I site must be at position 3.5.

Ear I has two cleavage sites, yielding fragments of 1.8 and 2.6 kbp. In the Ear I/Ase I double digest we see that the 1.8 kbp Ear I fragment is further cleaved to fragments of 0.7 and 1.1 kbp, leading to the conclusion that the two Ear I sites must surround the single Ase I site at 3.5 kbp such that one Ear I site is 0.7 kbp away and the other is 1.1 kbp away. The possibilities are

Table 6

Calculated Sizes of Restriction Fragments Plasmid pBR322

Gel 1[a]

Distance	kb	Distance	kb	Distance	kb
Lane 1: EcoR I		**Lane 2: Bal I**		**Lane 3: Ase I**	
9.3	4.4	9.3	4.4	9.3	4.4
Sum	4.4		4.4		4.4
Lane 4: EcoR I/Bal I		**Lane 5: EcoR I/Ase I**		**Lane 6: Bal I/Ase I**	
14.4	3.0	12.4	3.5	18.0	2.3
25.0	1.4	31.1	0.9	19.5	2.1
Sum	4.4		4.4		4.4
Lane 7: Ear I		**Lane 8: Ear I/Ase I**		**Lane 9: EcoR I/Ase I/Ear I**	
16.3	2.6	16.3	2.6	17.5	2.4
21.1	1.8	28.1	1.1	28.1	1.1
		34.2	0.7	34.2	0.7
				51.0	0.2
Sum	4.4		4.4		4.4

Gel 2[b]

Distance	kb	Distance	kb	Distance	kb
Lane 1: Drd I		**Lane 2: Drd I/Ase I**		**Lane 3: Drd I/Bal I**	
10.4	4.0	14.0	3.1	13.6	3.2
41.9	0.4	30.9	0.9	32.6	0.8
		41.9	0.4	41.9	0.4
Sum	4.4		4.4		4.4
Lane 4: Sno I		**Lane 5: Sno I/Bal I**		**Lane 6: Sno I/Ear I**	
15.7	2.7	21.1	1.8	16.9	2.5
27.0	1.2	27.0	1.2	27.0	1.2
38.5	0.5	30.9	0.9	41.9	0.4
		38.5	0.5	51.0	0.2
				61.0	0.1
Sum	4.4		4.4		4.4
Lane 7: Rsa I		**Lane 8: Rsa I/Drd I**		**Lane 9: Rsa I/Ear I**	
19.5	2.1	19.9	2.0	19.5	2.1
23.9	1.5	27.0	1.2	24.8	1.4
32.4	0.8	32.4	0.8	41.9	0.4
		45.8	0.3	60.9	0.1
		60.9	0.1		
Sum	4.4		4.4		4.0

[a] Slope = –31.3, intercept = 123.
[b] Slope = –31.4, intercept = 124.

$$3.5 - 0.7 = 2.8 \text{ and } 3.5 + 1.1 = 4.6 \text{ (i.e., 0.2)}$$

or

$$3.5 - 1.1 = 2.4 \text{ and } 3.5 + 0.7 = 4.2$$

The Ear I/Ase I/EcoR I triple digest resolves the uncertainty by cleaving the 2.6 kbp Ear I fragment to fragments of 0.2 and 2.4 kbp. Thus the EcoR I site, at 0.0 kbp, must be 0.2 kbp away from one Ear I site and 2.4 kbp away from the other. The Ear I sites, then, must be at 2.4 and 4.2 kbp.

Drd I also has two restriction sites, yielding fragments of 0.4 and 4.0 kbp. Further cleavage by Ase I (at position 3.5) converts the 4.0 kbp fragment to fragments of 0.9 and 3.1 kbp. Thus the Drd I sites are 0.9 kbp away from the Ase I site in one direction and 3.1 kbp away in the other, so that the possible locations of the Drd I sites are either at positions 2.2 and 2.6 kbp or at positions 0.0 and 0.4 kbp.

In the Drd I/Bal I double digest we see that Bal I, with a site at 1.4 kbp, cleaves the 4.0 kbp Drd I fragment to 0.8 and 3.2 kbp fragments, suggesting that the Drd I sites are at positions 0.2 and 0.6 kbp or at positions 2.2 and 2.6 kbp. Comparing the two double digests makes it clear that the data will be consistent only if we place the Drd I sites at 2.2 and 2.6 kbp.

There are three sites for Sno I; redigesting with Bal I cleaves the 2.7 kbp Sno I fragment to fragments of 0.9 and 1.8 kbp. We conclude that there are Sno I sites 0.9 and 1.8 kbp away from the Bal I site at 1.4 kbp. The possible Sno I sites are at positions 0.5 and 3.2 kbp or at positions 2.3 and 4.0 kbp.

When we examine the Ear I/Sno I double digest we note that Sno I cleaves the 1.8 kbp Ear I fragment to 0.2, 0.4, and 1.2 kbp and the 2.6 kbp Ear I fragment to 0.1 and 2.5 kbp. The 2.6 kbp Ear I fragment extends clockwise from position 4.2 to position 2.4, indicating a Sno I site either at position 4.3 kbp or at position 2.3 kbp. The latter is consistent with the Sno I/Bal I double digest data, so that two of the Sno I sites are at positions 2.3 and 4.0 kbp. Cleavage at these two sites yields the 2.7 kbp Sno I fragment. The remaining Sno I site is 0.5 kbp away from one of these sites, i.e., either at position 2.8 or at position 3.5 kbp.

If the last Sno I site were at position 3.5 kbp, then the Sno I/Ear I double digest would contain a 0.5 kbp fragment, which is not found. Only assigning the last Sno I site to position 2.8 kbp is consistent with the experimental data.

The Rsa I/Drd I double digest shows that Rsa I cleaves the 0.4 kbp Drd I fragment to fragments of 0.1 and 0.3 kbp, suggesting an Rsa I site at either 2.3 or 2.5 kbp.

The Rsa I/Ear I double digest is puzzling. The sum of the fragment lengths is 4.0 kbp, and only four fragments are shown. Since Drd I has two sites and Rsa I has three, one fragment appears to be missing in the double digest. If we understand that the "missing" fragment is 0.4 kbp in length, and that it comigrates with the other 0.4 kbp fragment, the mystery is cleared up.

Rsa I cleaves the 1.8 kbp Ear I fragment to 0.4 and 1.4 kbp, indicating an Rsa I

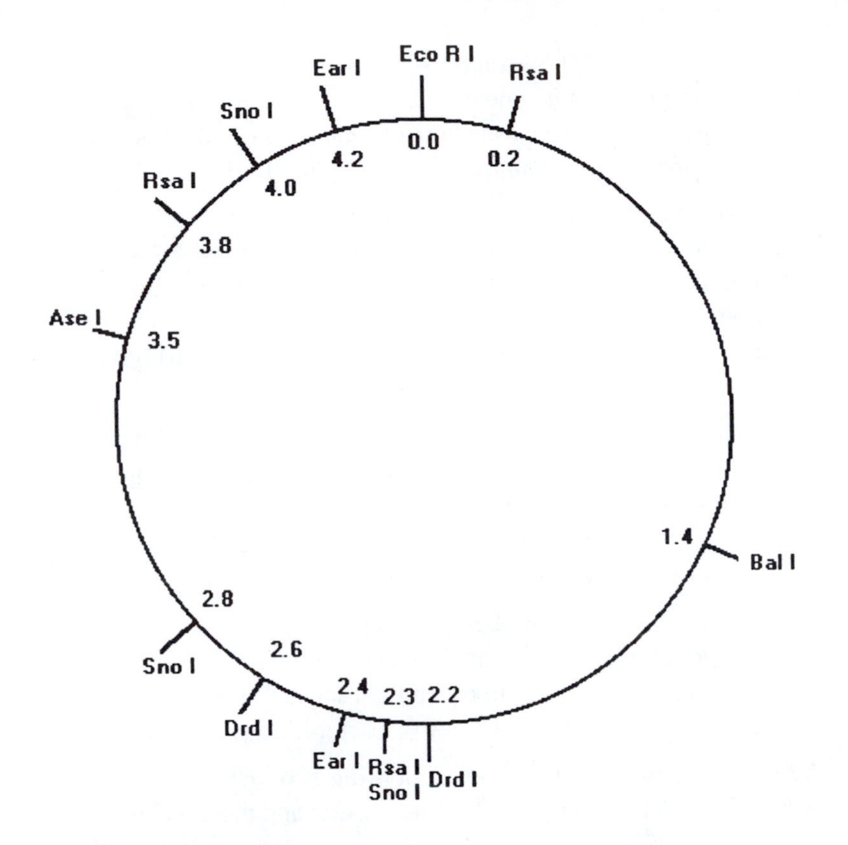

FIGURE 19 Restriction map of plasmid pBR322 (Problem 4.11).

site at either position 2.8 or 3.8 kbp. The possible sites, then, are 2.3 or 2.5 kbp, and 2.8 or 3.8 kbp. We can eliminate the 2.8 kbp site by realizing that a second Rsa I site at 2.3 kbp would yield a 0.5 kbp fragment in the single digest, and that a second Rsa I site at 2.5 kbp would yield a 0.3 kbp fragment; neither is found. Thus one of the Rsa I sites is at position 3.8 kbp. A second Rsa I site at 2.5 kbp would yield a fragment of 1.3 kbp, which is not found. Only an Rsa I site at 2.3 kbp yields the correct single digest data.

The third Rsa I site is either at position 0.2 or 1.5 kbp. A site at 1.5 kbp would yield a 1.7 kbp fragment in the Rsa I/Ear I double digest. Since this is not found, the third Rsa I site must be at 0.2 kbp.

A restriction map of pBR322 is shown in Figure 19.

Appendix B Answers to Additional Problems

Chapter 5

Problem 5.1

A = Lys–Leu–Lys–Asp–Gly–Phe–Ala–Glu

Problem 5.2

B = Pro–Val–Lys–Met–Tyr–Val–Val–Arg–Thr–Ser

Problem 5.3

C = Lys–Phe–Ser–Met–Tyr–Val–Arg–Leu–Phe–Thr

Problem 5.4

D = Glu–Ser–Arg–Tyr–Ser–Glu–Val–Met–Phe–Ala–Lys–Arg–Glu–Pro–Lys

Problem 5.5

E = Ile–Ala–Glu–Ser–Ala–Tyr–Val–Asp–Arg–Gly–Phe–Leu–Val–Arg–His–Ala–Phe (note that P1 contained 2 phenylalanine residues)

Problem 5.6

F = Ile–Arg–Glu–Ser–Tyr–Gly–Arg–Val–Arg–Pro–Leu–Ile–Phe–Cys–Leu–Ser–Asn

Problem 5.7

G = Phe–Ile–Val–Leu–Gln–Ala–Arg–Thr–Ala–Leu–Tyr–Asp–Gly–Arg–Ser–Tyr–Met–Thr

Problem 5.8

H = Thr–Lys–Glu–Thr–Trp–Pro–Lys–Ile–Arg–Pro–Phe–Met–Ile–Tyr–Cys–Leu–Ser–Gln

Problem 5.9

J = Leu–Met–Ser–Arg–Leu–Asp–Arg–Leu–Ser–Phe–Leu–Asp–Phe–Leu–Thr–Lys–Leu–Thr–Met–Tyr–Leu

Problem 5.10

K = Met–Gly–Arg–Ser–Tyr–Ser–Lys–Ser–Gly–Ala–Arg–Leu–Phe–Ser–Arg–Ser–Gly–Ala–Lys–Ser–Trp–Ser–Arg–Ser–Leu

Problem 5.11

The sequence of peptide L is shown in Figure 1.

L = Thr-Val-Glu-Lys-Gly-Gly-Lys-His-Lys-Thr-Gly-Pro-Asn-Leu-His-Gly-Leu-Phe-Gly-

Arg-Lys-Thr-Gly-Gln-Ala-Pro-Gly-Tyr-Ser-Tyr-Thr-Ala-Ala-Asn-Lys-Asn

FIGURE 1 Sequence of peptide L (Problem 5.11).

Problem 5.12

Peptide M is shown in Figure 2.

Problem 5.13

A. GCT GTC GAT CCC GTA GCT TAG CTG ATC GTA CGT CCC CGA TCG TGA TCG TTA TTT TTC GGC TAG CTT TTT

B. AAT TCG TCG GGC TTG GGG CTA GAA TGC TGT CGA TCG TAG TCG TAT TCG GGG CTA GCT AGT CGG ATC GAT

C. AAG CTT TAA TCG AAC GAT TTT AGA AGC CCA GCA CTA GCA TCA GCA TCA CGA TCA GCA TCG ACA CAG CAA

D. CGC GCG CCC TCG CCC TCA GCC TAC GAA AAA TCC ACG CCC TAC CGA TCC CCA GCA GCA CAG CAC CGG GGG

Problem 5.14

The restriction map is shown in Figure 3.

Problem 5.15

The complete λ restriction map, derived from the data in this problem, is summarized in Table 1. Note that the Apa I site was arbitrarily taken to be at 10.1 kbp. If it were assumed to be at 38.4 kbp instead, all subsequent assignments would result in a mirror image of the map given here.

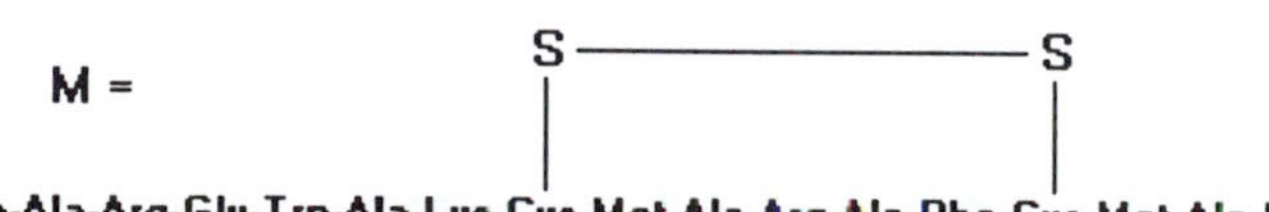

FIGURE 2 Sequence of peptide M (Problem 5.12).

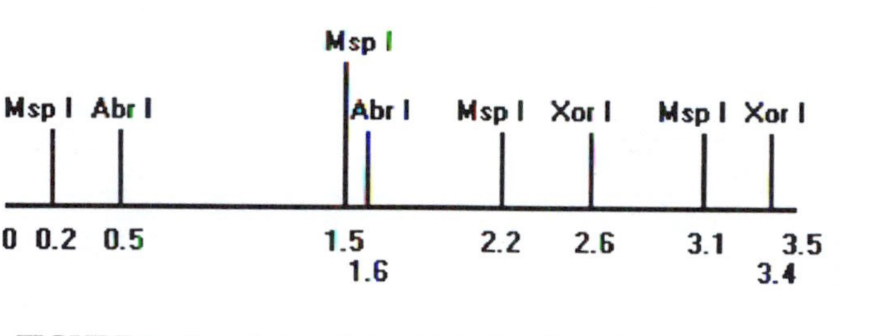

FIGURE 3 Restriction map of 3.5 kbp DNA (Problem 5.14).

Table 1

Restriction Sites in Bacteriophage λ DNA

Position	Enzyme	Position	Enzyme	Position	Enzyme
5.6	Sno I	19.9	Xma III	34.7	Nhe I
10.1	Apa I	21.0	Eco47 III	35.8	Pvu I
11.9	Pvu I	21.8	Sno I	36.7	Xma III
17.1	Kpn I	26.2	Pvu I	37.1	Eco47 III
18.6	Kpn I	27.1	Sno I	39.9	Sma I
19.4	Sma I	31.6	Sma I	40.2	Sno I

Index

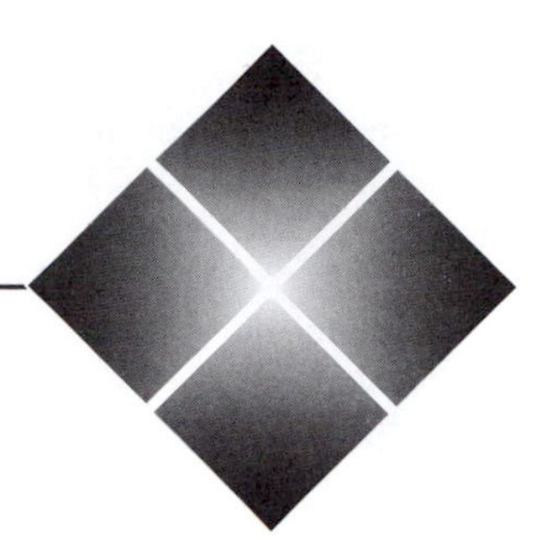

A

B

C

D

E

I

K

L

M

N

O

P

R

S

T

U

V

W

X